Workshop Electrics

Workshop Electrics

Alex Weiss

ARGUS BOOKS

Argus Books
Argus House
Boundary Way
Hemel Hempstead
Herts HP2 7ST
England

First published by Argus Books 1994

ISBN 1 85486 107 7

Phototypesetting by The Studio, Exeter
Printed and bound in Great Britain by Biddles Ltd., Guildford & King's Lynn

Contents

Chapter 1.   Introduction   1

Chapter 2.   Planning and preparation   3
Location and size of workshop, basic electrical principles, equipment to be
installed, power consumption, sources of electricity, assessing the household
installation, replacing consumer units, wiring regulations, planning and
positioning lights and power outlets, three phase machinery, inspection and
testing, sources of materials.

Chapter 3.   Safety in workshop electrics   15
The importance of safety, electricity and water, isolating the mains, wiring
colours, fuses and miniature circuit breakers (MCBs), extension leads and
adapters, rotating machinery and power failure, double insulation, plugs and
cables, battery chargers, working alone, first aid, fire risks, lightning, safety
rules.

Chapter 4.   Single phase supplies   26
Circuits, consumer units, standby power supplies, 110/120 volt equipment,
filters and suppressers.

Chapter 5.   Fuses and MCBs   33
Fuses, MCBs, residual current devices (RCDs), why fuses blow, repairing
fuses, checking for faults.

Chapter 6.   Wiring and connection to the mains   40
Permanent mains wiring, flexible wiring, cable stripping, permanent wiring
installation, surface and metal boxes, junction and terminal boxes, wiring up
different circuits, test equipment and testing, other types of wiring.

Chapter 7.   Outside workshops   49
Size and length of cable, underground and overhead connection, connection
at house and workshop, outside lights and sockets, security lights.

Chapter 8. Plugs and sockets   56

What to power from a 13 amp socket, spurs, installing sockets, 13 amp plugs and their wiring, flex connectors and extension leads, adapters, speed controllers.

Chapter 9. Fixed appliances   66

Fixed connection units, equipment requiring more than 13 amps, types of fixed appliance, heaters and thermostats, night storage heaters.

Chapter 10. Lighting   70

Circuits, junction boxes and wiring, switches, fluorescent and compact fluorescent lights, range of light bulbs, fixtures and fittings, task lighting, supplying lights from 13 amp sockets, inspection lights, emergency lighting.

Chapter 11. Three phase supplies   85

Advantages and disadvantages, obtaining a supply, three phase converters, running three phase machines, connecting to a three phase supply, wiring and distribution panels, starting and speed control.

Chapter 12. Low voltage supplies   90

Low voltage lighting, 12 volt equipment, power supply units, speed controllers, wiring and tools, telephones, intercoms, security systems, TV aerials.

Glossary of terms   98

List of useful addresses   102

Index   103

CHAPTER 1

Introduction

As with so many things, people are often frightened of electricity because they cannot see it and do not understand it. Equally, those who are familiar with the subject find little to worry about. The aim of this book is to help the amateur model engineer, who is not a trained electrician, to equip a small home workshop with all the electrical fittings needed. While there are many books available dealing with electricity in the home, no single volume contains all the information necessary for carrying out such work in a home workshop.

The editor of the magazine *Model Engineer* wrote in the early 1990s that there is only one real difference that his grandfather would have seen in the average home workshop of today. That difference is the replacement of belt drives and manual power with the ubiquitous electric motor. It is because of this change, allied to the need for good lighting, particularly during those long dark winter evenings, that this addition to the Workshop Practice series of books has been prepared.

Safety is a key issue with mains electricity, and it is most important that Chapter Three, which deals with this potentially lethal subject, is carefully read and digested before any work is undertaken. It is preceded by an explanation of the planning which should be undertaken before any electrical installation work commences. As most readers will be using 240 volt AC, single phase mains supplies, the vast majority of this book deals with the fusing, equipment, lighting, fixtures, fittings and wiring for this type of supply.

The needs of the modeller with an outside workshop have not been ignored as there are special rules involved in exterior wiring. Chapter Eleven is devoted to three phase supplies, which are a normal requirement for ex-industrial machine tools. It outlines how such machines may be powered in the home workshop using readily available single to three phase converters. The last chapter deals with low voltage supplies; well suited to providing lighting for machine tools as well as powering suds pumps and on tool electric motors. It also looks at security systems, telephones, intercoms and TV aerials.

At the end of the book, a glossary of terms helps to explain words, phrases and abbreviations which may be new to the reader, and a list of useful addresses of some sources of supply is also pro-

1

vided. Workshop Practice Series No. 16, *Electric Motors*, covers a specialisation purposely not dealt with in this volume.

A convention used in this book is that **words in bold type in the text apply to rules and regulations which must be obeyed**. There are a variety of regulations and legal requirements covering electrical installations in the United Kingdom, and all the advice given in this book conforms to latest practice at the time of publication in 1994.

There are some tasks which are considered, for one reason or another, to be unsuitable for the reader to undertake, or are beyond the scope of this book. These include the replacement of a single phase consumer unit and the installation of an Electricity Board three phase supply. In addition, although the advice in this book is generally applicable to home electrics, it does not cover the special aspects of electrical installations in bathrooms and kitchens, and **texts covering this particular subject must be consulted before any work is undertaken in such rooms**.

As this book is intended to be used as a reference, there is inevitably some repetition of information, to ensure, for example, that fusing information is available both in the chapter on fusing, and in the one on plugs and sockets. So, whether you are embarking on a brand new workshop building, or just planning to add an extra 13 amp socket to power a new tool, you should find the help you need within these pages.

CHAPTER 2

Planning and preparation

General

Who needs to plan! Let's get on and start the real work! This is generally a formula for disaster, and never more so than when dealing with electrical installations. As with any other activity in the home workshop, careful preparation will repay the model engineer many times over. The old adage of measure twice, cut once, means that a well planned set of circuits can save money, ensure a high degree of safety, and make future expansion much easier. So, let's start at the beginning.

Workshop location

The first question to be answered is where the workshop will be located. Will it be an indoor one, located in one of the rooms in the house, or is the loft to be converted? Alternatively, the garage may provide a more convenient location, and this may be attached to the house, or a completely separate building. Finally a shed in the garden is often the only acceptable solution, particularly if noise and dirt are expected by the owner, or even anticipated by the remainder of the family. For an external building without electrical power, special consideration will have to be given to the provision of a source of electricity to the building

itself, remembering that special rules apply to the type of cable, its installation and connection into the household supply.

The size of the workshop

It is not the role of this book to cover the various pros and cons of where and how to establish a workshop; only to deal with electrical matters. Thus, the next question is how large the workshop is to be (the area in square metres) and how high the ceiling will be. For many readers, however, the workshop will already exist, and their interest is solely in adding to an existing electrical system.

Basic electrical principles

This is the point at which some readers may run into difficulties due to a lack of understanding of the basic principles of electricity, and the next few paragraphs give a simple explanation. Those with a good understanding of the subject may, therefore, jump to the section in this chapter on workshop power consumption, and, if they are only planning some electrical additions to an existing workshop, move straight to Chapter Three.

Anyone who is going to undertake installation or maintenance of workshop mains electrical systems, or, for that

Water	Electrical	Electrical measurement units	Abbreviation
Pressure	Voltage	Volts	V
Rate of flow	Current	Amps	A
Size of pipe	Resistance	Ohms	R
Consumption rate	Power consumption	Watts	W
Amount consumed	Power consumed	Kilowatt hours	kW hr

matter household ones, will need a basic knowledge of electricity. At this stage, many people turn off, but, in truth, the necessary knowledge is really very simple. There are only six things which need serious consideration, and an analogy with water is useful in understanding these principles. We will look in turn at the electrical equivalents of the following familiar terms from the water analogy (see above).

Voltage
First, how much water pressure is there? The electrical equivalent is how high is the voltage. The mains voltage in the United Kingdom is 240 volts, while in almost all of continental Europe it is 220 volts. The third common standard is that used in the United States, where it is 110/120 volts. Compare this with an individual AA pencell, which produces 1.5 volts, and a car battery, which gives 12 volts. Basically, the higher the voltage, the more electricity can flow through a given wire, or conductor as it is known. Unfortunately, the higher the voltage the more potentially lethal an electric shock will prove to be. **Any voltage over 50 volts is considered potentially lethal**, and the next chapter covers the various safety aspects which need to be considered in dealing with electricity in the home workshop.

Current
Returning to the water analogy, how much water is flowing? In electrical terms, this flow, or electrical current as it is termed, is measured in ampères (normally shortened to amps). The average workshop appliance will require a few amps; a car starter motor in the region of fifty amps, and a battery powered transistor radio, a few hundredths of an amp. It is worth noting that the flow of current through a conductor causes a heating effect.

Resistance
The size of the electrical conductor, or wire, is just like the size of the water pipe, and obviously, the higher the electrical current flow in amps, the larger the cross section of the conductor needed. Here, there is another safety issue, since trying to pass too much electrical current through too small a conductor can easily lead to overheating and fire. This resistance to flow is measured in Ohms, and for conductors, is typically some tenths or hundredths of an ohm per metre. Insulation is normally measured in millions of ohms, abbreviated to megohms.

Power consumption
Next, we need to consider how much electrical power is being consumed, the equivalent of the rate of water flow, and this is obtained by multiplying the volts by the amps to give watts; the basic unit of electrical power. Thus a car starter motor may use 50 amps × 12 volts = 600 watts, whilst a lathe may well consume 6 amps × 240 volts = 1440 watts,

or 1.44 kilowatts, to use the measure preferred by the electricity supply industry. The conversion from watts to horsepower (HP) is simply done by dividing the wattage by 760.

Power consumed

Multiply the electrical power consumed in kilowatts by the duration of usage, and we arrive at the number of kilowatt hours for which we have to pay. This is an exact analogy to the amount of water used. The price for one kilowatt hour, one unit as it is known, is widely quoted, and can vary from area to area. It can also be dependent on the total annual usage as well as the time when the electricity is used.

Direct and alternating current (DC & AC)

The last electrical matter to be learned is the difference between alternating and direct current. Direct current, or DC, is very much what we would expect, the current flows from one end of the wire to the other, from, say one battery terminal, through a bulb, and back to the other terminal of the battery. In fact, this is exactly what happens. For reasons associated with electrical generation and transmission, it is much more economical to use so called alternating current, or AC, for mains electricity, where both the voltage and current alternate in a sine wave form, oscillating fifty times per second, or at fifty Hertz in the UK and Europe. In the USA, the frequency is sixty Hertz.

Three phase supplies

As far as the home workshop is concerned, single phase alternating current is the norm, but in fact, electricity is generated with three phases, and each house is normally fed with a single phase; thus the name. Industrial premises, which consume far greater amounts of electricity than the average home workshop requires, commonly use three phase supplies, and there can be a need for such an electrical source, even in a home workshop. In this case, there are three live wires, one for each phase, a neutral and earth. The electricity supplied is at 415 volts. Such a supply is most likely to be needed where ex-industrial machine tools have been acquired, and Chapter Eleven deals with this subject in some detail.

Electrical equipment to be installed

Moving on, it is important to decide what electrical equipment is likely to be installed in the workshop during its life. In particular, the more power thirsty items, such as lathes and milling machines should be identified. Fortunately their physical size makes them of prime importance during the general layout planning of a workshop. Clearly, the workshop with just a vertical pillar drill is going to have very different electrical needs to one equipped with a full range of machine tools. Always allow for future expansion, with additional sockets and fixed connection units; even a spare high power circuit if the purchase of a heavy machine tool is envisaged at some later date.

The following checklist may help decide what equipment will require electrical power in the workshop:

Lathe	*Suds pump*
Second lathe	*Electric welder*
Vertical milling machine	*Air compressor*
Horizontal milling machine	*Hot glue gun*
Shaper	*Soldering iron*
Jig borer	*12 V battery charger*
Pillar drill	*Nicad battery charger*
Tool cutter and grinder	*TV/Radio*
Surface grinder	*Electric kettle*
Bench grinder	*Vacuum cleaner*
Rotary polisher	*Dust buster*

Power hacksaw
Bandsaw
Circular saw
Fret saw
Scroll saw
Radial arm saw
Portable drill
Angle grinder
Electric planer
Abrasive band machine/
  linisher
Orbital sander
Belt sander
Disc sander

Heater
Security alarm
12 volt power supply
Router
Dust extractor
Ventilation fan
Hot air gun
Pressure washer
Electric stapler
Electric screw driver
Electric paint stripper
Heat shrink gun
Electronic test
  equipment

Workshop power consumption

Having identified the list of items, the next thing to do is to discover their electrical power consumption in watts. This information is usually given on a plate attached to the machine, and should always be quoted in manufacturer's brochures. First, are there any items of equipment requiring more than three kilowatts (nominally 4HP)? If so, each such tool will require a special high power circuit. Any other piece of equipment, with a lower power requirement, can be fed from a ring main.

Do not forget that the workshop may have to be heated by electricity. This type of heating has the advantage, from the workshop point of view, of providing moisture free warmth. Fan heaters can be plugged into a convenient socket, whereas wall-mounted heaters should be wired as fixed appliances. Night storage heaters require a completely separate low tariff electrical supply, with its associated timing and distribution systems, and should only be considered if such a supply already exists in the house.

Calculating a workshop's total electrical power requirement becomes particularly important if a new electrical installation is being planned, and especially if the workshop is to be located in a self-

contained outside building. It is probably a good idea to be able to consider the power requirements both in terms of watts and kilowatts; the normal units of power, and also in amps, the units which decide the size of electrical conductors needed.

A single lighting circuit can operate 12 lights, or a maximum of 1200 watts, whichever is the smaller. This equates to 12 one hundred watt bulbs, or eight 150 watt ones. Unless the workshop lighting is to be powered from an existing lighting circuit, which is already near to capacity, this limitation is unlikely to cause any difficulty.

As far as electrical sockets and fixed appliances are concerned, there is no limit to the number of sockets which may be fitted to a single ring main circuit, but **no single device may consume more than 13 amps or 3.12 kilowatts. There is also a floorspace limit of 100 square metres per circuit**, which may be a problem if the workshop is connected into an existing house ring main circuit. There are also limitations on the number of spurs which may be connected to a ring circuit. Basically, whilst **the same number of spurs as sockets is permitted, only one spur may be run from any individual connection to the ring main.**

Where a piece of equipment draws more than 13 amps, then 15, 20, 30 or 45 amp circuits must be established from the consumer unit to feed it, providing 3.6kW, 4.8kW, 7.2kW and 10.8kW respectively; the latter figures obtained by multiplying the current by 240 volts. For the purposes of this book, it is considered so unlikely that machines needing more than 10.8 kW of single phase power will be installed, that no further consideration is given to such a situation. A more detailed calculation is needed where a cable is to be installed to provide the total power to an outside workshop. In this case, a five amp circuit for lighting plus a further 30 amp circuit for a ring main will normally suffice, giving a total of 35 amps. To this must be added any high current circuit for any individual piece of machinery. Normally, a 45 amp circuit from the household consumer unit will be more than sufficient, providing 10.8 kilowatts of total lighting and power, on the basis that not all machinery will be operated at the same time.

Sources of electricity
The assumption has been made that electricity for the workshop will be obtained from the normal domestic supply which already exists in the owner's house. Other possibilities include the use of an off-peak circuit to power night storage heaters, and the possibility of a three phase supply. Consideration may also have to be given to the use of a petrol or diesel electric generator at remote sites.

Assessing the household electrical installation
It is important to make an assessment of the present household electrical installation before any major work to extend it into the workshop commences. Start by locating and looking in the 'meter

Fig 2.1 *The inside of a typical meter cupboard showing the Electricity Board's junction box and meter on the left, and the consumer unit on the right. The two cables connecting the consumer unit to the meter show up clearly.*

7

cupboard'. Has it got a consumer unit or is it so ancient that it still has old fuse and switch boxes? Houses which have not been rewired since 1947 will have a radial power circuit with round pin 5 amp and 15 amp plugs and sockets. If the electrical system is of this vintage, then the house needs rewiring, and **no attempt should be made by the amateur to extend such a system into a home workshop.**

Assuming a consumer unit has been fitted, the condition of the unit and cables running to and from it should be checked, including the state of the earth connection. The consumer unit will be fitted either with fuses or miniature circuit breakers (MCBs), it doesn't matter which. Any circuits protected by residual current devices (RCDs) should be noted for possible use in the workshop. Next check that the circuits are correctly rated and labelled, and, if this is not the case, make the necessary changes and complete the labelling. Details of how to do this are given in Chapter Five. Finally, see if there are any spare circuits available for powering the workshop, or assess whether the workshop can be powered from existing lighting and ring circuits. If a low tariff supply has been installed, a meter with two consumption indicators will have been fitted, as well as an Electricity Board time clock to switch between normal and low cost tariff.

Replacing the consumer unit

It is possible that the consumer unit may need replacing with a larger one to provide the necessary additional circuits. Whilst this job can be undertaken by the competent amateur, there are several reasons why the task is considered beyond the scope of this book. The consumer unit has to be disconnected by the Electricity Board, and the electrical installation checked and reconnected by them.

Furthermore, the whole house will be without electrical power whilst the unit is being changed. It is, therefore, a task best left to a professional electrician.

Wiring regulations

The Regulations for Electrical Installations or *IEE Wiring Regulations*, Sixteenth Edition, published in 1991, is the bible for all workshop electrical work. A dauntingly large document filled with legal phraseology, it is available directly from the Institution of Electrical Engineers or from public libraries. The first edition was published as long ago as 1882! The regulations are designed to protect persons and property from the hazards of electric shock, fire, burns and injury from mechanical movement of electrically actuated machinery, the latter where it can be

Fig 2.2 The Regulations for Electrical Installations, *published by the Institute of Electrical Engineering provide the basis of the rules for workshop electrical work in the UK.*

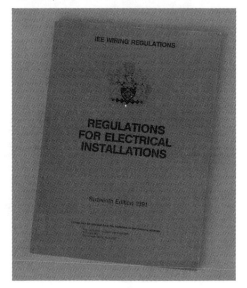

prevented by electrical emergency switching or by electrical switching for mechanical maintenance. Whilst the regulations are non-statutory, they may be used in court in evidence to claim compliance with a statutory requirement. As well as safety, they cover the selection and installation of equipment, as well as inspection and testing. They also quote the relevant British Standards and statutory regulations. **The latest edition has standardised on the term Phase conductor to define what was until that time called the Live conductor**. This book uses the term phase throughout, but it should be noted that some accessories may still carry the letter L adjacent to the phase terminal.

Having said that, there is no need to rush out and purchase a copy of the regulations. This book incorporates the basic rules and regulations which relate to workshop electrics at the time of going to print. It is worth realising that there are summary guides to the *IEE Wiring Regulations* available, which present the rules in concise and more comprehensible form. Even the IEE themselves have published three booklets, called guidance notes to the sixteenth edition *Wiring Regulations*. The titles are *Selection and Erection*, *Isolation and Switching*, and finally *Inspection and Testing*. They are not just a summary or overview of the regulations, but do provide some amplification of the regulations.

Planning the number of lights and power outlets

A fair amount of thought is required when planning the electrical supplies for a new workshop. The two key rules are to ensure that there is plenty of light, and that there are too many, rather than too few power sockets. It is surprising how many electrically powered devices are to be found in the modern home workshop, and these, together with portable lights, will all demand 13 amp sockets or fixed connection units. The plan should not be based on the use of multi way adapters, despite the fact that these may be needed at some later time.

Each item of permanently installed equipment will require its own fixed connection unit located within two metres run of cable. Although more than one tool may be powered from a single 13 amp socket by plugging and unplugging the devices, for preference, most are likely to be left permanently plugged in, but switched off at the socket. It is almost as cheap to fit double sockets as single ones, if the cable cost is included, so these should normally be chosen. As to how many sockets, a minimum of ten and preferably twenty are good figures for starters. A typical list, which will obviously depend on the results from the checklist already prepared of electrically powered workshop equipment, might read as follows:

Fixed appliances

Lathe	*Milling machine*
Vertical drill press	*Power hacksaw*
Bench grinder	

13 amp sockets

Soldering iron	*Light for lathe*
Fan heater	*Light for vertical*
Light for milling machine	*drill press*
Light for bench grinder	*Light over vice*
12 volt power supply	

Positioning switches and sockets

Light switches should normally be located 1.4 metres above the floor, and sockets certainly **no less than 150 mm above the floor**. They should preferably be just below or 150 mm above workbench level. They should be well distributed around the workshop to avoid the dangers of trailing

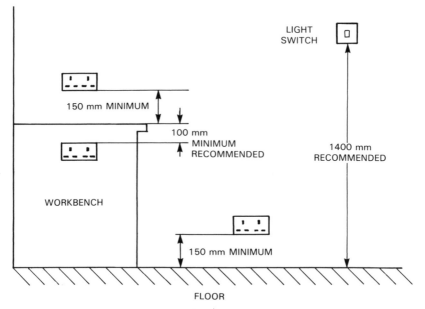

LIGHT SWITCH

150 mm MINIMUM

100 mm MINIMUM RECOMMENDED

1400 mm RECOMMENDED

WORKBENCH

150 mm MINIMUM

FLOOR

Fig 2.3 *The required vertical position of 13 amp sockets does ensure not only that they are safely placed, but also that they are easy to reach. Light switches 1.4 metres above floor level come easily to hand.*

flexes. The need for off-peak electricity, three phase supplies and even low voltage power needs should be considered at this stage.

Three phase machinery
For those with machinery requiring three phase supplies, the possibility of getting the local Electricity Board to provide such a power source may be considered, but the difficulties of the installation work put it beyond the scope of this book. There is the alternative of a single to three phase converter, which can be particularly cost effective if only 1 or 2 three phase machines are to be used. The main factors to consider are the number of machines, the number, power rating and starting current of their

electric motors, and whether more than one motor will be run at any one time, as well as the capacity and space required for a converter.

Lighting the workshop
Moving to the important subject of illumination in the workshop, it is essential to consider what type of lighting will be fitted. There are two types of light which are required. The first is background illumination, which provides a useful amount of lighting for the whole workshop, and which is switched on when anyone enters the room. Fluorescent lighting is ideal for this purpose, providing economic, shadow-free illumination. Sufficient incandescent lamps can, of course, achieve the same function, but

10

with significantly higher running costs.

The second type of illumination is task lighting. This is light which increases the level of illumination in a particular work area, or on an individual machine tool. Generally, task lighting needs to be directional, as well as aimable, so that the position of the pool of light can be adjusted to suit the worker. Spotlights and angle poise lamps are ideal for this rôle.

Communications to the workshop

Where the workshop is located remotely from the house, an intercom to the house can be a real advantage in maintaining harmony in the home. The ability of anyone in the house to communicate with the workshop, without either trailing outside, or shouting, gives benefits which can far outweigh the initial cost.

The addition of a telephone extension can be particularly advantageous when the occupant of the workshop is the only person at home.

Workshop security

In a time when high crime rates are a sad fact of life, the value of the equipment in a home workshop is often far greater than realised. A security alarm can give peace of mind, as well as acting as a deterrent to would-be burglars. Its installation is straightforward if it is merely an addition to an existing household system, but rather more complex if undertaken from scratch.

The tools required

Most of the tools and equipment needed for electrical installation and repair work should be found in the tool box of the

Fig 2.4 *A selection of tools for electrical work, including two types of wire strippers, pliers, wire cutters, large and small screwdrivers and a sharp knife.*

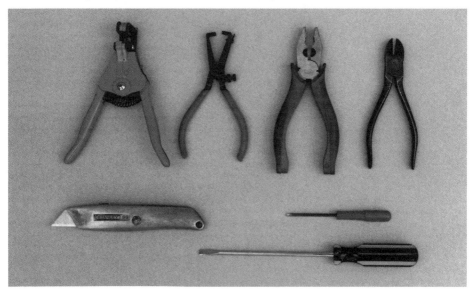

average model engineer. In fact, only a very small number of tools are required for electrical work itself, but a much larger number are necessary when laying cables, or installing electrical fixtures and fittings.

The basic tools include small electrical screwdrivers, wire cutters, wire strippers, and a sharp knife. Perhaps less commonly found are a torch, a mains detector, which can be in the form of a special screwdriver with a neon in the transparent plastic head and an insulated metal shaft, or a small battery operated, hand held unit with a small red LED which glows, and a buzzer which bleeps when the unit is held close to any live cable connected to the mains. A continuity tester is another useful device, and one can be constructed simply from a battery, buzzer or bulb and bulb holder, three short lengths of insulated wire and a pair of crocodile clips. Some multimeters include a continuity testing facility, and such items are well worth considering as the basic tools of the electronics and electrical enthusiast.

Normal DIY tools will be needed for such tasks as lifting floorboards and cutting away plaster and brickwork, as well as making good when the work is done. Thus hammers, cold chisels, saws, plastering knives, drills and masonry bits will all have their uses. It is assumed that anyone planning to build, or who already has a workshop, will be familiar with the use of these tools and the tasks to be done. If not, any good book on DIY should be consulted.

Fig 2.5 *Items of test equipment likely to be found in the home workshop include digital and analogue multimeters, a home made and commercially available continuity tester and a mains live wire detector, the last built from a simple kit available from Maplin.*

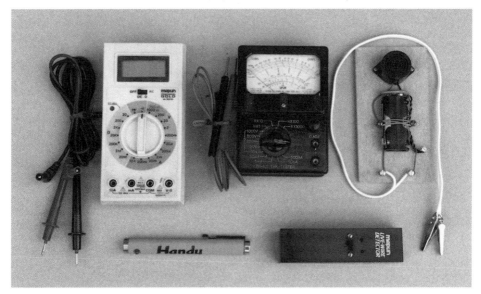

Standards of workmanship

Because safety is involved, a high standard of workmanship is of paramount importance. Sticking to the rules and wiring diagrams given in this book is obviously important, as is **ensuring that all screw connections are firmly tightened and covers replaced**. Fortunately, most model engineers are used to following instructions, and the use of machine tools, power saws and steam boilers all require major attention to safety. Consideration of electrical work in the same light will go a long way towards ensuring a good quality of installation.

Electrical interference

Electrical interference from workshops is something well worth thinking about these days. Not only can electrical interference affect neighbours' radio and television, it can do the same in the workshop owner's home. Fortunately, the problem is generally not a common one; interference from the brush gear of portable power tools being the main culprit. Motor suppresser capacitors for small electric motors are both cheap and readily available. Suppression units are also available for three phase motors, should it be found that these are causing problems.

Inspection and testing

Checking and **testing a new electrical system before it is finally connected to the mains is clearly a necessity**, and such tests are always carried out by the Electricity Board before any consumer unit is connected to their supplies. For many, pre-testing is better carried out by a professional electrician, and, in any case, has advantages if it is carried out by a third party. Regardless of what changes have been made, be they alterations or additions, they must comply with *IEE*

Fig 2.6 *A 13 amp plug checker is simple and quick to use, as well as being virtually indestructible. It will show whether or not a socket is correctly wired, and clearly identify the problem.* (Photo courtesy Martindale Electric Ltd.)

Wiring Regulations, and must not impair the safety of any existing installation. The checks needed are comprehensive, and require the use of specialist test equipment, some of which is unlikely to be available to the model engineer. Therefore, all new electrical work should be tested by an electrician who is registered with the National Inspection Council for Electrical Installation Contracting, before any connection to the mains is made.

Sources of materials

The first rule to remember, when purchasing materials for electrical installations, is that in general, you get what you pay for. Some foreign imports are still less than acceptable from a quality point of view, so please ensure that purchases are made from a reputable supplier. These concerns do not apply to cable for wiring the installation, as all

13

cables sold in the UK have to meet the relevant British Standards. The two main sources, both of materials and cabling, are DIY superstores, and specialist electrical suppliers; the latter often providing comprehensive illustrated catalogues. The big advantage of the former for the amateur is the ability to browse around and see exactly what is available, and at what price, before committing to a purchase. A good catalogue can, however, provide almost as much help.

Some less commonly used items of equipment will require the location of specialist suppliers. Good examples include the devices needed to operate three phase machinery from a single phase supply and low voltage safety lighting. A list of useful addresses is give at the end of this book.

CHAPTER 3

Safety in workshop electrics

The importance of safety
It is difficult to write about electrical safety without lecturing the reader and giving the impression of potential danger. However, providing sensible precautions are taken, the risks are negligible. The fundamental danger of working with electrical systems is that of fatal electrocution, and such deaths occur with alarming regularity in the UK each year. Basically, allowing mains electricity to flow through the human body causes, at best an extremely unpleasant shock, and at worst, the heart to stop working, with fatal consequences. Electric shocks can also cause very unpleasant burns, usually deep but localised to the point of contact. **It is thus most important that this chapter is read and fully digested before any electrical task is undertaken in the workshop.**

Electricity and water
The dangers of electricity are significantly increased by the presence of water, as tap water is an excellent conductor. Thus, there are special rules for bathrooms, kitchens and other places where running water is to be found in the house, and these rules mean that it would be very difficult to install workshop tools safely in such a room.. As a result, **no home workshop should be located in any room with running water**. The importance of earthing as a way of providing an alternative path to the human body in case of a faulty appliance cannot be over-emphasised, and the workshop owner must be fastidious in the connection of earth leads where required.

Isolating the mains
It is common sense to turn off the electricity before work of any sort is undertaken on any circuit. Perhaps less obvious is the need to ensure that no one else turns it back on while work is proceeding. If all electrical power in the house needs to be shut off, then a clear notice attached to the isolating switch on the consumer unit warning others not to turn it back on is the best idea. It should be used in conjunction with a verbal warning to all other members of the household.

Where only the circuits to the workshop require isolation, then the removal of the relevant fuses or miniature circuit breakers in the consumer unit, having turned off the isolating switch first, is the right approach. Again verbal and written warnings must be used; these being par-

ticularly important if the workshop shares power with any other rooms in the house. As a further safety check that the circuit being worked on is safe, a small electrical screwdriver, with a neon light in the handle can be touched on the wires to ensure that they are dead. **Any appliance which requires repair must be unplugged from the mains socket before any work commences. If the appliance is connected by a fused connection unit, then that unit must be switched off, and its internal fuse removed**. A further check should then be made to ensure that the electrical wiring to the appliance is dead.

Wiring colours

Sadly, there are two different sets of colours used for the single phase mains wiring found in home workshops. **For permanently installed wiring, RED is used for PHASE (formerly known as LIVE), BLACK is used for NEUTRAL and YELLOW/GREEN for EARTH. For flexible wiring connecting an appliance, or tool, PHASE is BROWN, NEUTRAL is BLUE and EARTH is again YELLOW/GREEN. The correctly coloured wires must always be connected securely to the right terminals**. It is also worth noting that many electrical fittings are still annotated with an L for connection to the LIVE wire, although, as mentioned above, this nomenclature has recently been changed to PHASE. Details of three phase wiring are given in Chapter Eleven.

As has already been said, the danger of overloading circuits is that the wiring will overheat and cause a fire. By its very nature, much electrical wiring runs under floorboards or along ceiling joists, and when the typical oily workshop environment is added, a potentially highly inflammable situation exists. As well as the rules governing the number of lights that can be run from a single lighting circuit and the floor area that can be supplied from a single ring main circuit, **any tool or heater drawing more than 13 amps (3.12kW) must be supplied from a special high power circuit**. Do not under any circumstances be tempted to overload any electrical circuit by running either too many electrical devices from it or by running a device requiring more power than the circuit is safely capable of giving. **No tool or item of machinery must ever be run from a lighting circuit**, although portable lights may be run from a ring main circuit.

Fuses and MCBs

Fuses and miniature circuit breakers (MCBs) are designed to disconnect a circuit or appliance in the presence of an excessive flow of electrical current. They are available in a wide variety of current ratings, and it is important that the

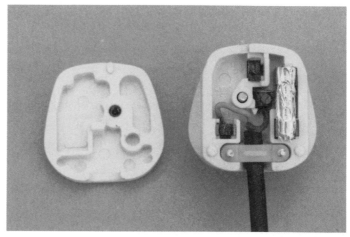

Fig 3.2 *The wrapping of a failed fuse in metal foil as a quick fix, when a replacement fuse is not available, is a dangerous practice which is to be deplored.*

correctly rated fuse or MCB should be fitted to each circuit and each plug or fixed connection unit. The deplorable habit of **wrapping a piece of silver foil around a failed fuse, to provide a temporary repair in the absence of a replacement fuse, is a highly dangerous practice**, as it completely negates the function of the fuse, and can easily lead to burned out insulation of the wiring at best and a fire at worst. **The use of old nails or anything but the proper fuse is equally risky**.

RCDs

Residual Current Devices (RCDs), and their predecessors, Earth Leakage Circuit Breakers (ELCBs), are designed to reduce the risk of electrocution significantly. Their use is becoming generally much more widespread and is essential where there is a risk of the tool cutting the cable, as for example with an electric lawnmower or hedge cutters. Their use in the home workshop is recommended, and **is compulsory for sockets supplying outdoor garden tools or equipment**. A

unit with a trip current of 30ma is ideal for use with power tools and machines, lower ratings giving unnecessary problems of false trips with some workshop

Fig 3.3 *If an RCD is not fitted to the circuit at the consumer unit, or built into the socket, a plug in RCD is a simple and effective way of providing safety from electrocution.*

equipment. **Power tools and other elec-trical equipment should never be used out of doors except in dry weather conditions**.

Extension leads and adapters

There is a particular problem associated with the type of extension lead which can be wound on and off a reel. **These leads are designed to be used only when all the wire is unwound from the reel**. The danger is of overheating and fire as, when wound on the coil, the wire will be denied its essential air cooling. There are occasions when flexible wiring proves to be too short, and a semi permanent joint becomes essential. A proper flex con-nector must be used, or, even better, the complete lead replaced with a single longer piece of flex. A multi way adapter, employed sensibly, is a useful device where insufficient 13 amp sockets exist. It is, however, a temptation to overload the particular socket, which still has an electrical limit of 13 amps. This tempt-ation must be resisted at all costs.

Rotating machinery

Mains cables and rotating machinery are a dangerous enough combination without adding to the problem with trailing mains flexes close by. The consequences of such a flex getting caught up in the moving parts of any machine tool are

Fig 3.4 *A coiled up extension lead is likely to cause the cable to overheat, and possibly lead to a fire in the coil. Always unwind the lead fully before use. Note that it is no longer permissible to locate sockets so close to floor level, though many such installations still exist.*

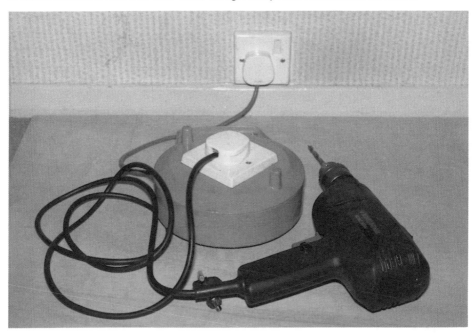

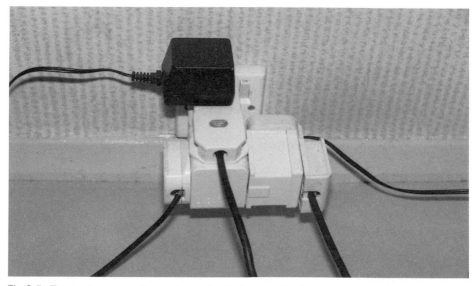

Fig 3.5 *The socket shown is totally overloaded, as are the first two of the adapters. It is quite unacceptable to use three adapters in series, with five devices being fed from the single socket. This most dangerous practice could easily cause an electrical fire.*

very severe. It is quite possible for the metal of the machine itself to become live, and assuming an operator is working with the machine, touching it could easily be fatal. Thus, great care must be taken to keep all cables clear of machinery, and, as an added precaution, low voltage motors and lighting should seriously be considered for 'on machine' applications.

Power failure and rotating machines

When there is a power failure to a machine tool, for whatever reason, if the tool is not turned off, then when the electrical supply is restored, the machine will immediately start operating again. This is clearly a dangerous situation for anyone in the workshop, but special switches are available which automatically turn off when the power fails. Their use is

highly recommended for the control of any machine tool. Fortunately, most modern machines are now fitted with such switches.

Double insulation

The majority of portable power tools are double insulated. This means that all the electrical components are contained within a plastic housing, which is usually the body of the tool itself. Double insulated tools do not require an earth lead back to the mains, but do still require the correct fuse within the 13 amp plug. It is important to recognise, however, that a number of tools are still not double insulated, particularly older ones. **For these tools, a three lead cable, with a properly made earth connection is essential**.

Fig 3.6 *The two push buttons, labelled 0 and 1, and coloured red and green, ensure that the machine is automatically switched off in the case of mains failure.*

Plugs and cables

A plug must always be fitted to any piece of portable electrical equipment. It is essential the wires are connected to the correct terminals and that the flex to the plug is securely held by the anchor device built into the plug. Details of how to do this are given in Chapter Eight. The plug itself should never be removed from its socket by pulling it out by the cable. In fact, cables should be inspected regularly for damage, and anchor clips

Fig 3.7 Never attempt to pull out a plug by its cable. It overstresses the anchor point in the plug, and can quite easily lead to a short circuit within the plug itself.

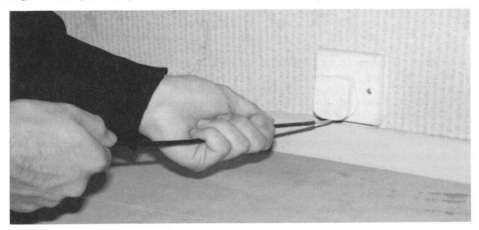

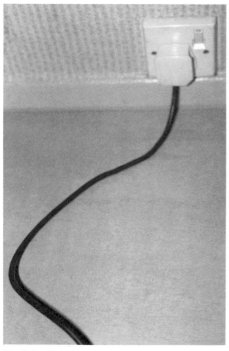

Fig 3.8 *A trailing cable across the workshop floor is an invitation for someone to trip over it and get hurt. The remedy is to fit sufficient sockets around the workshop to obviate the problem.*

Fig 3.9 *A taut cable will stress the anchor point in the plug in the same way as trying to pull out the plug by its cable.*

checked for effective operation. Flex should never be pulled taut, allowed to trail across floors, nor be run under carpets. Fit rubberised plugs where rough usage is likely, and this is particularly so with portable tools. Where a flex is too short, either replace it with a longer piece, or use a proper flex connector. **Never ever try to insert bare wires directly into a socket.** Always fit a correctly fused plug. **Finally, do not, under any circumstances, twist the wires together and cover the joint with insulating tape.** It just isn't safe.

Battery chargers

The hazards of charging lead acid batteries may not seem to be an electrical problem, though the charger is normally run from the mains. The risk is that hydrogen is generated during the recharging process, and that this gas, combined with the oxygen naturally occurring in the atmosphere can easily be ignited to cause a serious explosion. The spark caused when a light is turned on or the charger is turned off is quite sufficient to ignite the hydrogen, causing devastating damage. Such battery chargers should only be operated in well ventilated areas. Likewise, nickel cadmium batteries which are fast charged for too long have also been known to explode, firing nasty shrapnel considerable distances. Fast chargers with built in timers or charge level sensing devices are the only sensible solution, but still need to be used with care and in accordance with the manufacturer's instructions.

21

Fig 3.10 *Trying to connect anything to a socket by the bare wires is a very dangerous practice, which can easily give the person attempting such a thing a potentially lethal electric shock.*

Fig 3.11 *Never, ever, be tempted to join flex by twisting the wires together and binding them with insulating tape. A proper flex connector must be fitted, with the female socket connected to the 13 amp plug, and the male socket to the appliance.*

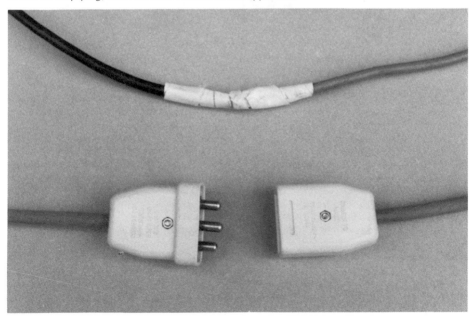

Working alone

Working by oneself in a home workshop always carries the danger of injury, with no-one aware of the accident and thus available to give first aid. This is a doubly dangerous situation when working with electricity, as immediate action needed in cases of electrocution is to disconnect the person from the source of electricity. It is for this reason that, whenever practical, electrical installation and maintenance work should only take place when another adult is available and aware both of the work being carried out and the action to be taken in the event of an accident.

First aid

Should the worst occur, it is important to understand the basic rules of first aid as they apply in the case of electric shock. It is sensible for anyone owning a home workshop to have attended first aid classes. If someone gets an electric shock, and is still in contact with the source of the shock, the first action must be to **switch off the electricity at the nearest switch and/or pull out the plug**. It may even be necessary to go to the consumer unit to turn off the power. If this cannot be done, the victim must not be touched or the current may flow through the rescuer as well. Using a dry cloth or piece of clothing, the victim should be pulled clear of the source of the electrical shock. Alternatively, the person can be knocked clear of the electrical device using a piece of insulated material, such as a length of wood.

Once disconnected from the mains electricity, do not try to move anyone who has fallen, as they may have sustained other injuries. Cover them with a blanket or coat to keep them warm until they have recovered. Medical advice should be sought if any injury is at

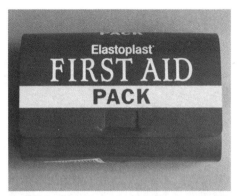

Fig 3.12 *A well equipped first aid kit should be kept in the workshop, not just to deal with electrical burns, but any other minor injuries which may occur.*

all serious. If in doubt, dial 999 to summon emergency help. In severe cases of electric shock, the person involved can stop breathing, and artificial respiration will be needed; a skill taught to first aiders.

Fire risks

The location and construction of the workshop will have a considerable impact on safety considerations. Wooden workshops present a much greater fire risk than masonry ones, and those in or attached to the house increase the risk of any fire spreading to the house. All appliances should be turned off at the mains when not in use, and plugs removed from their sockets where practicable. Only those fire hazards which are of an electrical nature are considered here.

Electrical fires mainly result either from overloaded circuits or faulty insulation. Adherence to the advice given in this book will minimise these risks. However it is clearly sensible to fit one of the modern self contained smoke alarms in the workshop as the loud warning these emit will provide early warning of a fire.

Fig 3.13 *A powder-filled fire extinguisher can be used on any electrical fire, and should be conveniently located in the workshop. Smoke alarms are a low cost way of getting early warning that there is a fire in the workshop, and can help to save not only lives, but also valuable equipment and models.*

This will provide added safety to occupants and give the opportunity to extinguish the fire before serious damage is done to the house, workshop or its equipment.

Every home workshop should include a fire extinguisher, and one suitable for fighting electrical fires is essential (not a water based one). Most modern home extinguishers use powder to smother the fire, and these are ideal. The dangers of fighting an electrical fire with water should already be clear to the reader, due to the normally excellent conductive characteristics of water.

Lightning conductors
Lightning usually strikes the highest object, tree or building in a particular area. The fitting of a lightning conductor

is therefore normally only a consideration if a workshop is isolated. It is certainly worth thinking about, although few home workshops are likely to need such a safety feature.

Conclusion
Finally, if there is ever any doubt about what is the correct electrical procedure, or how some task should be done, always ask the advice of a qualified electrician. It's better to be safe than sorry.

Summary of safety rules
- *In case of electric shock*
  - *Turn off or otherwise remove the source of electrical power.*
  - *If necessary, apply artificial respiration.*
  - *If the victim has fallen to the floor, do not try to move them, due to possible*

other injuries.
- Cover the victim with a blanket to keep them warm.
- Seek medical help.
- Before undertaking electrical work, turn off power, affix a notice, and tell others.
- Test that the power is off before starting work.
- Before working on an appliance, remove its plug from the socket.
- If the appliance runs from a fixed connection unit, turn off the electricity and remove the fuse.
- Always use a correctly fused plug to connect workshop tools.
- Fit rubberised plugs where rough usage is likely.
- Join flex only by using special flex connectors.
- Never try to insert bare wires directly into a socket.
- Trailing flexes are dangerous and can cause injury.
- Don't overstretch flexes or try to pull a plug out of a socket by its flex.
- Always power equipment used out of doors through an RCD.
- Always use cable and flex of the correct current rating.
- Ensure that the correct rating of fuse is always fitted.
- Always locate and repair a fault before replacing a fuse or switching on an MCB.
- If in doubt, consult a qualified electrician.

CHAPTER 4

Single phase supplies

General

As indicated in Chapter Two, 240 volt single phase AC electricity is the normal alternating current provided by the Electricity Boards to homes in the United Kingdom. It is not generated or transmitted by them in this form, for practical and economic reasons. The high tension cables and their associated pylons may run at up to 240,000 volts, but this voltage is locally transformed to the required level for household use. The electrical supply enters each house through a service cable, via a fuse, meter and possibly a junction box, all of which are owned by the Electricity Board. **No attempt must be made to tamper with them**. They are in fact secured with lead seals and it is a serious offence to break such seals. The Boards publish tariff rates, which indicate the price of their electricity. This is quoted as a price per unit, or kilowatt hour; that is the amount of electricity used by a one kilowatt device being run for one hour, or a hundred watt bulb being run for ten hours.

The consumer unit is fed from the meter or junction box by household owned cables, **which may only be connected to the meter or junction box by the Electricity Board** when the installation is commis-

sioned. In a few older installations, a number of switch boxes are fed from the Electricity Board junction box. As mentioned earlier, it is recommended that these houses are rewired with a consumer unit since the age of their wiring is such that it is considered unsafe to connect up additional loads to them. Such rewiring is beyond the scope of this book, and should be carried out by a professional electrician.

The only other items likely to be found in the meter cupboard (or more recently on the integral garage wall) are the time clock and distribution unit for low tariff off peak supplies to night storage heaters and immersion heaters. In this case, the meter will include two indicators; one for low tariff and the other for normally priced electricity.

Circuits

Lighting, ring main and circuits supplying more than 13 amps (15, 20, 30 or 45 amps) are all supplied from the consumer unit, which also incorporates a switch to allow the electricity to be isolated from all the circuits. In fact, a ring main uses 30 amp conductors and a similar rating of fuse, on the basis that not more than this current will normally be drawn from

Fig 4.1 *This meter cupboard contains a time clock and a twin dial meter, which switches charging to low tariff overnight. The physical condition of this installation leaves something to be desired.*

all the fixed connection units and sockets in the ring. A typical three bedroom semi will have two lighting circuits; one for upstairs and one for down and two ring main circuits, likewise for upstairs and down, and probably a 30 amp electric cooker circuit. Hopefully, there will also be a spare unused fuseway to allow for expansion at a later date.

Consumer units and connecting to them

A new consumer unit may be needed for a number of reasons, but for the purposes of this book, only two are considered. The first is the case where there are

insufficient spare fuseways, or none at all available to allow electricity to be extended to the home workshop. The second is the case of the outdoor workshop, where a small consumer unit can provide a safe and tidy distribution unit within the workshop.

Changing a consumer unit is quite an awkward task, since it requires the Electricity Board to disconnect the consumer unit and reconnect it after installation of the new unit. During the intervening period, the home is totally without electricity. Furthermore, the Electricity Board is entitled to refuse to reconnect power unless the complete household electrical installation meets its safety standards. It is thus not a task that should normally be undertaken by an amateur. Wiring a consumer unit into a new home workshop in a self contained building, on the other hand, is a straightforward task and is covered in detail in Chapter Seven.

Where an existing consumer unit has sufficient spare fuseways, a check should be made of the existing fuseways, checking they have the correctly rated fuses and MCBs; 5 amps for lighting circuits

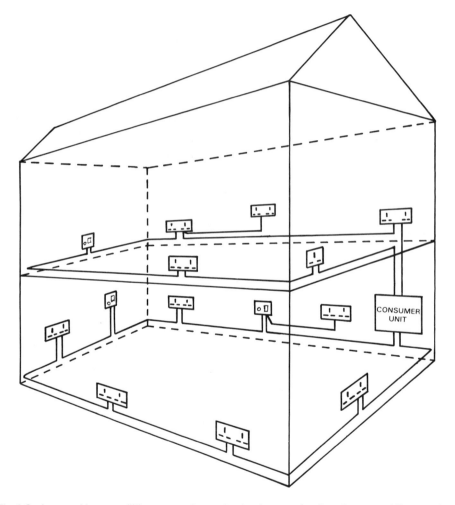

Fig 4.2 *A normal house will have two ring main circuits; one feeding the ground floor sockets and fixed connection units, and the other the upstairs ones.*

and 30 amps for ring main circuits. The circuits should also be correctly labelled as, for example, upstairs ring main and downstairs lighting.

New circuits for the workshop are connected to the consumer unit once all the wiring has been completed and tested. The electricity must be turned off at the consumer unit and the cables fed neatly behind the unit. The neutral and earth wires are connected to the relevant common terminal strips, having been cut

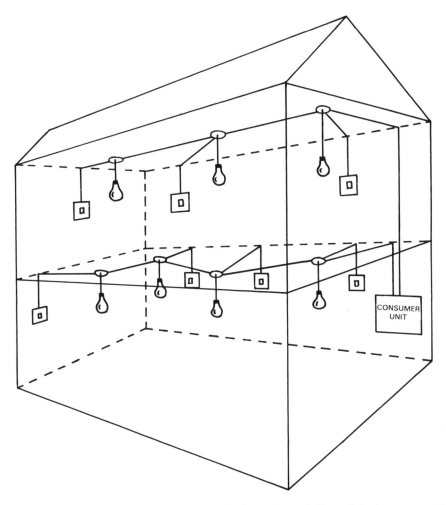

Fig 4.3 *Two lighting circuits are normally provided, one for each floor of the house, providing radial connection to each light and its associated switch.*

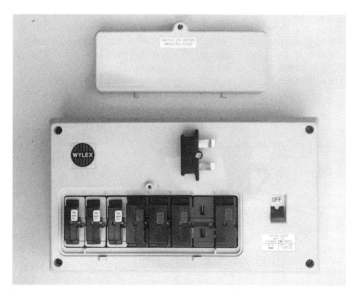

Fig 4.4 *A typical consumer unit. The cover for the fuses has been removed. This one has eight ways and has been fitted with cartridge fuses.*

Fig 4.5 *The most modern type of consumer unit, this split unit is fitted with MCB protected units, some of which are also RCD protected, and some of which are not.* (Photo courtesy MK Electric.)

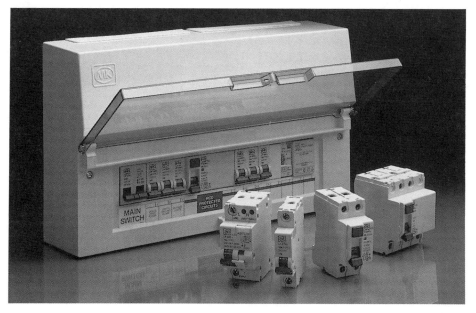

to length. The phase wire, or pair of wires in the case of a ring main circuit, is then connected to the terminal associated with the selected spare fuseway. Finally, a fuse or MCB of the correct rating is fitted and the power restored.

Standby power supplies

The Electricity Boards are not the only source of electrical power. Petrol or diesel driven generators and battery powered inverters are to be found in many homes as a back up source of power in the case of mains failure, or as portable power supplies. Whilst they are ideal for providing a small amount of power for such items as a central heating pump, television, or a small number of lights, they have little application to the provision of electrical power in home workshops.

US 110/120 volt equipment

Ever since the Americans came to the UK en masse during the Second World War, there has been a sprinkling of ex US equipment requiring 110/120 volts for operation. **On no account must such equipment be connected directly to the UK mains**. What is needed is a transformer, of a suitable rating, to reduce the 240 volts to the American 110/120 volts. Such transformers, whilst expensive and bulky, are often significantly cheaper than purchasing a new tool to replace the American one, which are generally available in the UK at very low prices. However, only equipment which can satisfactorily be run at fifty Hertz is suitable. Since the US standard is sixty Hertz, this means that synchronous motors, ones where their speed is controlled by the frequency of the mains, will run approximately seventeen percent slow. Thus a motor designed to operate at 3,600 rpm will only turn at 3,000 rpm. It is for the workshop owner to decide if this speed reduction is acceptable.

Such a transformer, wired into the mains via a 13 amp plug or a fixed appliance unit, can then be provided with an output socket to allow the US standard tools to be plugged in. **Under no circumstances use 13 amp plugs and sockets for this task due to the danger of inadvertently plugging the tool into the UK mains**. US pattern flat pin plugs and sockets are readily available in the UK from component suppliers such as Electromail and Maplin.

Filters and suppressers

One of the problems which can beset the family and neighbours of home workshop owners is the problem of mains-borne interference. This problem usually manifests itself in the form of picture interference and noise on the sound channel of the television, sound noise on the radio and problems with the home computer. It is most commonly produced by brush motors and speed controllers. For those who run into this problem, there are two approaches to solving the difficulty; tackling at source and tackling it at the equipment experiencing the problem. The former is the right approach for the home workshop owner, and the fitting of suppression capacitors (small electrical storage devices) is a simple and inexpensive task. With a delta configuration, the capacitors should be connected across the two motor terminals as well as to earth.

Summary

It can thus be seen that for most home workshop applications, the provision of 240 volt single phase electrical power is not only a normal, straightforward task, but will also meet most of the electrical needs of the average workshop. Chapter Eleven gives details of three phase

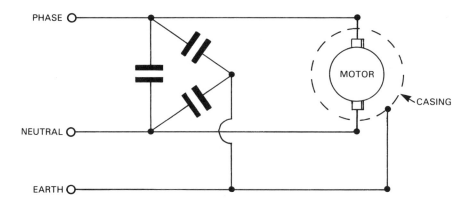

Fig 4.6 *Mains electric motors fitted with brushes are a common cause of electrical interference, but can easily be suppressed by using three capacitors in a delta configuration.*

supplies, while Chapter Twelve provides information about low voltage power systems.

CHAPTER 5

Fuses and MCBs

General

Either a fuse or a miniature circuit breaker (MCB) of the correct rating is required in every circuit. This is to prevent the wiring overheating, and causing a fire, in the event of circuit overload, due either to human failing or to a wiring or equipment fault. A fuse can be regarded as the weakest link in the system, which fails before damage is done to any circuit or piece of equipment. Fuses are to be found in the line supplied by the Electricity Board, in the consumer unit, in fixed connection units, in 13 amp plugs, adapters and extension leads, and even within some electrical equipment itself. **Electricity Board fuses must never be touched**, and in the unlikely event of a failure of this fuse, a call must be made to the Board's emergency service to get the fuse replaced. Failure to provide adequate fusing can lead to serious fires occurring, as well as damage to wiring and equipment. MCBs are increasingly to be found in consumer units in place of and providing the same function as fuses. The use of residual current detectors (RCDs) is a major contributor to providing safety from electric shock, and **their use is mandatory when operating equipment out of doors**.

Fuses

The fuses encountered by those dealing with home workshop electrics are normally only of two sorts; those found in consumer units, and those found in 13 amp plugs, adapters and extension leads. The ones used in plugs are in the form of throwaway cartridges, and once they have blown must be replaced with another of the correct rating. Similar, but larger cartridge fuses are used in consumer units, and exactly the same rules

Fig 5.1 *5 and 30 amp consumer unit cartridge fuses and their carriers, showing how the carrier is easily dissembled with a screwdriver to allow fuse replacement.*

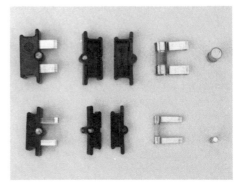

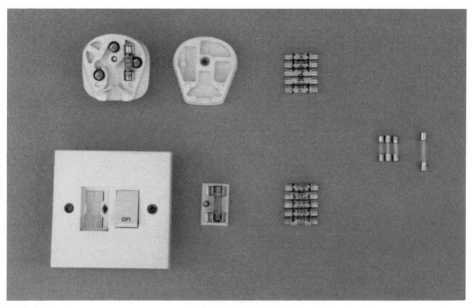

Fig 5.2 *A 13 amp plug with a selection of fuses and below, a fixed connection unit with its fuse carrier and identical fuses to those used in plugs. To the right is a selection of glass equipment fuses.*

apply as far as replacement is concerned. As an alternative in consumer units, replaceable wire fuses may still be found, where repair is undertaken by manually replacing the length of burned out fuse wire with a new length of the correct rating.

Selecting the right fuse to fit in a 13 amp plug is a straightforward task, the fuse supplied normally being a 13 amp one, which may well be of far too high a current capacity. The right fuse should be chosen to match the current rating of the device which it supplies. If the only figure available is quoted in watts or kilowatts, these should be divided by 240 or 0.24 respectively to arrive at the current consumption. The next highest fuse rating should then be selected, but

the manufacturer's instructions regarding fuse rating should always be followed.

Some items of electrical equipment carry internal fuses for additional protection. These are usually small glass cartridges, either 20 millimetres or occasionally 1¼ inches long. They come in a wide range of ratings from 50 milliamps to fifteen amps, and both quick blow and anti surge variants are available. Their location within the equipment varies from the easily accessible fuses in some light dimmers to the much less accessible ones in the back of television sets. **(Beware of taking the back off a television set. It can give an electrical shock from its high voltage power supplies to the tube, even after the set has been disconnected from the mains.)** It is a simple task to see visually when the fuse has blown, and it may be replaced with one of identical type and rating. **Always disconnect equipment from mains electrical power before removing any covers to access internal fuses.**

Fuses come in a variety of ratings and colours, and the lists below detail those used in consumer units as well as those designed to fit in three pin electrical plugs.

Consumer unit fuses

Current	Colour	Use
5 amp	*White*	*Lighting circuits*
15 amp	*Blue*	*Immersion heaters and single 13 amp or 15 amp sockets*
20 amp	*Yellow*	*Night storage heaters*
30 amp	*Red*	*Ring circuits and machine tools up to 7.2kW*
45 amp	*Green*	*Machine tools between 7.2 and 10.8kW*

13 amp plug cartridge fuses

Current	Colour	Use
2 amp	*Black*	*Portable lights, radios and TVs*
3 amp	*Red*	*Portable lights, radios and TVs, low power workshop tools and appliances up to 720 watts capacity.*
5 amp	*Black*	*Small tools of up to one kilowatt consumption*
10 amp	*Black*	*Medium powered appliances up to two kilowatts*
13 amp	*Brown*	*Most motor-driven equipment and electric heaters*

2, 5 and 10 amp fuses are less commonly used than 3 and 13 amp ones.

Miniature Circuit Breakers (MCBs)

Increasingly, MCBs are becoming the popular way of protecting circuits, because of the ease with which they can be turned off and reset. They are designed to fit into modern consumer units and they are rated by current, in the same way as fuses, but incorporate a

Fig 5.3 *A pair of Miniature Circuit Breakers (MCBs), which have been removed from a consumer unit. The one on the left has tripped, shown by the protruding button.*

button or a switch. The button allows the MCB to be turned off, in exactly the same way as will occur in the event of an overload. Thus it is obvious, in the event of a failure, which MCB has been tripped due to the position of its button or switch. The circuit is simply restored by switching on the MCB again. If it will not stay in the on position, **the fault, which still exists in the circuit, must be rectified before further attempts are made to reset the MCB.**

Residual Current Devices (RCDs)

RCDs are designed to protect people from electric shock and electrocution. They operate by detecting even the smallest current flow from the phase wire likely to cause an electric shock, and automatically tripping an inbuilt circuit breaker, fast enough to protect the person who has caused the trip. RCDs have now replaced the old Earth Leakage Circuit Breakers (ELCBs), though

Fig 5.4 *A variety of 13 amp sockets, all incorporating a Residual Current Device (RCD). An RCD must always be used when powering equipment to be used out of doors.* (Photo courtesy MK Electric.)

reference to and use of the latter may still occur.

Some RCDs will trip in the event of a temporary power failure, thus preventing machinery from suddenly being energised when the power is restored. Ideally, from a safety point of view, all items of workshop equipment should be supplied through an RCD. It is recognised that this may be quite an expensive proposition, particularly if the existing consumer unit will not accommodate RCDs, but the expenditure should be considered very seriously in view of the considerable safety benefits.

There are three forms of RCD likely to be of use in the home workshop. The first is the small RCD which fits into a consumer unit, and protects a particular circuit or circuits. The second type is built into a 13 amp socket; the standard double sized unit in fact housing a single socket in addition to the RCD. The third type looks like an adapter, plugs into any socket, and allows any item of equipment drawing 13 amps or less to be plugged into it. **An RCD must be employed for any socket feeding garden tools or outside equipment**. This will include any tool which is taken outside the workshop for use, such as an electric welder or air compressor.

RCDs are fitted with a test circuit, which, by use of a test button, checks the safe operation of the RCD. This test function should be checked each time a piece of equipment is to be operated out of doors. RCBOs are now available, which combine the functions of MCBs and RCDs and are able to distinguish whether the trip was caused by an overload situation or a leakage to earth.

Why fuses blow

Fuses are the safety links in electrical circuits, and are designed to prevent overheating and fires being caused by electrical overloads or short circuits. Basic fuses, including cartridge fuses, incorporate a short length of thin wire which melts under overload conditions, breaking the circuit. **A fuse of a lower current rating than the wiring it is protecting must be fitted to every circuit.**

MCBs sense the overload and automatically switch off the circuit, allowing the MCB to be reset once the fault has been removed. MCBs are obviously more expensive than fuses, but the convenience of being easily reset has made them increasingly popular. Fuses for consumer units may be in the form of a plug-in holder, which can be rewired with fuse wire by the home owner, or the holder may utilise replaceable cartridges. The more modern consumer units have been designed for MCBs. The fuses in 13 amp plugs, adapters, extension leads and fused connection units are always of the cartridge type.

Very occasionally, a fuse will fail for no apparent reason apart from age, and a simple replacement will restore the electricity. Usually, however, a fuse blows because there is a short circuit somewhere, caused by a problem with the equipment or its wiring. These faults may range from a loose wire in the plug to a fault in the equipment itself. Faults in the household wiring itself are, fortunately, extremely rare. The more common causes of a fuse blowing are:

● *Too many items of equipment being operated from a single socket, using adapters.*
● *Too many powerful bulbs being used in a single lighting circuit.*
● *A faulty connection or wiring within a plug.*
● *A fuse with too low a rating being fitted.*
● *Worn or damaged flex or circuit cable to a piece of electrical equipment.*
● *An internal equipment fault.*

It is important that any fault is discovered and rectified before the fuse is replaced or the MCB switched back on. Failure to do so will not only overload the wiring, but also lead to the fuse re-blowing.

How to repair fuses

Before attempting to change a fuse in a consumer unit, **the isolating switch on the unit must be turned off.** Do not be tempted to ignore this rule just because there are a number of digital clocks powered by the circuit, which will need resetting. The offending fuse holder may then be removed and the wire or cartridge replaced with a similar item of the same current rating. In the case of fusewire, the burned out wire will be obvious, and can be replaced with a new length from a card which carries spare 5, 15 and 30 amp fusewire. In the case of a cartridge, with no visible sign of failure, it will be a case of seeing if the change of cartridge has affected the repair. A better alter-

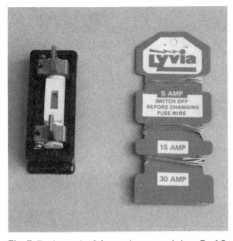

Fig 5.5 *A card of fusewire containing 5, 10, and 30 amp rated wire, together with the type of consumer unit fuse which utilises such wire.*

native is to check the cartridge with a multimeter set to ohms or continuity, or to use one of the low cost continuity testers now available.

As with a blown fuse in a consumer unit, when a fuse in a plug blows, **the plug, its wiring and the equipment it is powering must be checked to correct the fault which caused the original failure.** To change the fuse, **the plug must first be removed from the socket.** Then undo the central screw(s) on the underside which holds the plug together, and lever out the fuse with a small electrical screwdriver, reversing the tasks when the replacement fuse has been fitted. It is worth remembering that if a fused adapter or extension lead is being employed, it may well be that fuse which has blown. With a fixed connection unit for a wired in appliance, the procedure for fuse changing is similar, although in this case, **the unit's switch must be turned off.** The fuse is accessed through a small panel in the front cover of the connection unit, and again, is most easily levered out using a small screwdriver.

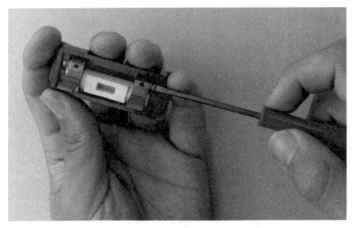

Fig 5.6 *Changing the wire in a consumer unit fuse requires the wire to be fed through the hole running through the central ceramic part and wound around the retaining screws, which are then tightened.*

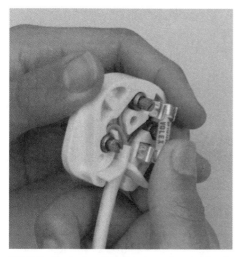

Fig 5.7 *Replacing the fuse in a plug is simply a matter of removing the cover, levering out the failed fuse, pushing in a new one of the correct rating and replacing the cover.*

Checking for faults

The importance of locating a fault before a new fuse is fitted is paramount. Unfortunately, faults can often be difficult to locate, and a methodical system of fault finding will save time and effort. If a fuse fails in a consumer unit, start by establishing whether the failure occurred when a light, or piece of equipment was switched on, or whether there was no apparent cause for the failure. In the former case, a check of the offending item will usually lead to the cause of the fault. In the latter case, removal of part of the electrical load, followed by replacement of the fuse, will quickly tell if the failure was a simple overload situation. Remember that the most common cause of a fuse blowing in a consumer unit is that one of its circuits has been overloaded by trying to run too many items from it.

Fault rectification in an item of equipment can be more difficult, and **no investigation should start until the particular item has been disconnected from the mains.** The nose can be a very helpful sensor for fault location. The smell of overheated insulation is easy to detect and location of the source straightforward. Thus, the problem should be dealt with straight away, even if dinner is waiting on the table! The most common reason for a fuse blowing in a ring main supplied item of equipment is due to a fault in the plug or the flexible cable to the equipment.

For rotating machinery, the motor itself is a prime suspect, and a check for free mechanical rotation is a good starting point. The motor can often be heard to hum, even though no rotation is taking place, In oscillating power tools, a faulty bearing can easily cause a motor to overload. Finally, a continuity check can indicate whether the fault is caused by a short circuit to earth.

CHAPTER 6

Wiring and connection to the mains

General

There are two basic types of wiring used in home workshops. The first is fixed wiring, employing a single strand conductor (occasionally more than one strand) sheathed in PVC. It is used for permanent house wiring. It is flexible enough to be bent into position during installation, but thereafter, will never normally be moved. The second type has a number of strands forming each conductor, and is flexible enough to sustain a fair amount of movement and bending; thus the name flex. This type of cable is used to connect up portable devices such as power tools and moveable lights.

Within each of these two classes of wiring, cables are available with different current ratings to provide for varying power needs, and different types of insulation depending on the particular use for which they have been designed. In the UK, all cables have to meet the relevant British Standard, so that any normal source of supply will provide a satisfactory product.

Permanent mains wiring

The standard cable used for permanent house and workshop wiring has two PVC insulated conductors, **RED for PHASE** and **BLACK for NEUTRAL**, and an uninsulated earth conductor; all three being contained in a white or grey PVC sheath. In the highest current ratings, multi strand, single insulated conductors are more normal. There are six common sizes of cable cross section, most of which are likely to be needed in wiring a home workshop from scratch. The chart shows the conductor size required for each of the types of circuit encountered in the home workshop.

Wire sizes for permanent circuits

Current rating	Conductor size	Use
5 amps	1.0 sq mm	Lighting
15 amps	2.5 sq mm	15 amp radial circuits/spurs
20 amps	2.5 sq mm	20 amp radial circuits/night storage heaters
30 amps	2.5 sq mm	Ring main circuits
30 amps	4.0 sq mm	30 amp radial circuit
45 amps	6.0 sq mm	45 amp radial circuit
60 amps	16.0 sq mm	Connecting consumer unit to meter

Exceptions to the standard three conductor cables are the four core cables needed for two-way light switches and the cables required for three phase

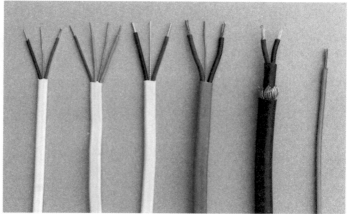

Fig 6.1 *A selection of cables showing from left to right: lighting cable, two way lighting cable, ring main cable, 45 amp cable, armoured cable and earthing cable.*

supplies. These latter may have three or four insulated conductors and an earth; the actual number depending on the requirement for a neutral conductor.

The cable used for running power underground to outside workshops has a very different insulation and protection system. There are two main specialist types. The first, designed for underground use, is an armoured cable and is available with two or three conductors and the armoured sheath. Each conductor is individually insulated with PVC, and similar insulation is provided inside and outside the protective sheath of steel wire. Alternatively a mineral insulated, copper sheathed (MICS) cable is available, where the PVC covered conductors are packed in magnesium oxide within a copper outer sheath, itself covered in PVC. Both these cables are expensive, and require junction boxes to connect them to normal interior quality cable. However, for many applications, normal cables can be used above ground for outside connection to buildings, and details of these are given in Chapter Seven.

Flexible wiring

Flexible wiring for appliances has, as already mentioned, a number of strands of wire in each conductor to give it the necessary flexibility, without fatigue failures causing the conductors to break. Again, it is available in a number of current ratings, and flex of an appropriate capacity (cross sectional area) must always be selected.

Wire sizes for flexible wiring

Current rating	Conductor size	Use
3 amps	0.5 sq mm	Lights up to 100 watts
5 amps	0.75 sq mm	Lights and appliances up to 1 kW
10 amps	1.0 sq mm	Appliances up to 2 kW
13 amps	1.25 sq mm	Appliances up to 3 kW
15 amps	1.5 sq mm	Appliances up to 3.5 kW
20 amps	2.5 sq mm	Appliances up to 4.5 kW
25 amps	4.0 sq mm	Appliances up to 6 kW

Where the cable is to be used out of doors, as for example with an electric welder, then a bright orange cable outer is recommended, to provide greater

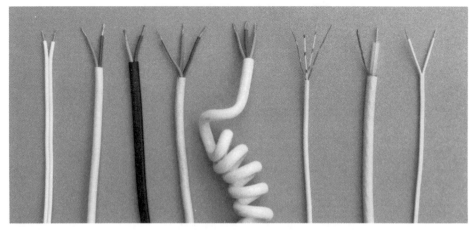

Fig 6.2 *A wide choice of flex is available including, from left to right: twin core flex, two core sheathed flex, two core flex sheathed in a black outer, three core sheathed flex, coiled flex, telephone wire, co-axial cable and bell wire.*

visibility because of the potentially more vulnerable position in which such cable is often placed.

Cable stripping

Removing the insulation from cables is an awkward, rather than a difficult task, and should always be undertaken with a proper wire stripping tool. These come in a variety of designs and prices, and are generally easy to use. The outer sheath of the cable will first need to be cut back, taking great care not to cut any of the insulation of the individual conductors. The wire strippers are then set to the correct wire diameter and used to remove the required length of insulation from each of the conductors. Where flex is being stripped, the conductor wires should then be twisted to form a neat ending.

Permanent wiring installation

Permanent house wiring can be laid in a number of places; wiring beneath the floor, wiring above the ceiling, or buried in the walls. In addition, it is permissible to run PVC-insulated and sheathed cable on the surface or within surface conduits. Although surface mounting is somewhat unsightly, it may be acceptable in a workshop environment, and does avoid

Fig 6.3 *Cutting away the sheathing with a sharp knife needs care to avoid damaging the insulation surrounding each wire.*

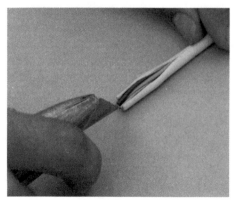

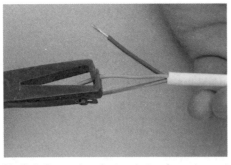

Fig 6.4 *Removing the insulation from the end of each wire is a straightforward task once the stripper has been set to the correct diameter for the wire.*

the problems of sinking the cables into the walls.

Underfloor wiring

Starting with wiring beneath the floor, this requires the lifting of at least some of the floorboards. It is not a very practical proposition where the floor covering is large sheets of floor quality chipboard. Where solid concrete floors have been installed, channels can be cut in the concrete, conduits fitted, and the channels re-concreted. For preference, a different route for the wiring should be found to avoid such hard work. Lifting floorboards is not a difficult task, but requires some forethought to ensure a minimum number are lifted. They can be eased up at one end using a brick bolster or wide wood chisel, and, if full length, cut using a tenon saw so that only a short length of the board is removed. The cable is simple to lay if it can be run parallel to the joists. If the run is at right angles to the joists, it will require a 12.5 mm hole to be drilled in each, about 50 mm from its top, to allow the cable to be fed through easily.

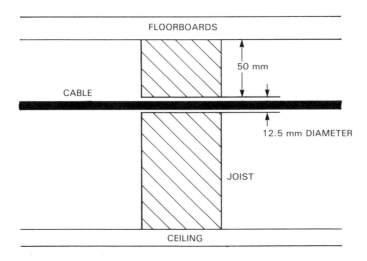

Fig 6.5 *When passing cables through joists, a 12.5 mm hole should be drilled, 50 mm from the top of the joist, and the cable threaded through.*

Wiring in the roof

Where cable is being run through a loft, it may be laid over the joists, assuming there is no intention to board over the loft floor at a later date. Two other important factors need consideration. The first is the need to run the cables over, rather than under, any roof insulation, particularly if high current radial circuit wiring is involved. The second is to run the cable through holes made in the joists in the area around the water storage tank and the entrance to the loft, where people are likely to need to stand on the joists. Where expanded polystyrene roof insulation is used, it is better to avoid any contact between the insulation and PVC-sheathed wiring. This is due to a chemical reaction between the two substances, which, though not dangerous, produces a substance on the outer cover of the cabling, which makes it appear that the cable covering has cracked if the cable is moved.

Wiring within the walls

The technique used for working on walls very much depends on the type of wall construction. Plastered brick/insulation block can have a channel cut in the plaster and the wiring clipped in place, with no further protection necessary, prior to being plastered over. However, a range of plastic conduits is available, which can be plastered into the wall, and which do allow the cabling to be replaced at a later date. Even simpler is to attach the wiring directly to the wall, with cable clips spaced at 400 mm intervals. Whilst not very attractive, it is a quick, cheap and practical solution for a home workshop. It is an important rule that buried cables in walls should only be run vertically to avoid the dangers of clashing with items being fixed to the walls at some later date. Cables may be run horizontally, but this should be avoided whenever possible, and if done, only in the top 150 mm or bottom 300 mm of

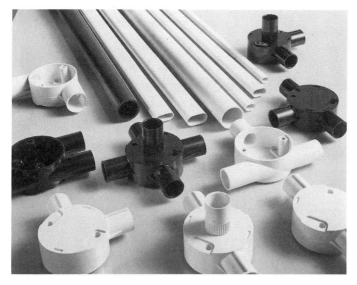

Fig 6.6 *A wide choice of conduit is available on the market, including handy plastic conduit, which is simple to cut and install.* (Photo courtesy MK Electric.)

Fig 6.7 *Cable which has been fixed to the wall with cable clips may well be the quickest and cheapest solution for a home workshop.*

the wall. **Cables must never be run diagonally across a wall**.

Hollow walls can be more problematic. The vast majority of these walls will be plaster board on timber studding. Locating a gap between the timber verticals, the cable can be threaded through holes drilled through the wall plate and any cross members, with an opening cut for the light switch or power socket.

Surface boxes and metal boxes

When installing light switches, sockets or fixed connectors, the two choices are to surface mount them or to flush mount them. Surface mounting requires a plastic box, which screws to the wall and houses the wiring and rear portions of the unit. The appropriate plastic webs will need to be broken out to allow cable entry. The surface box is screwed to the

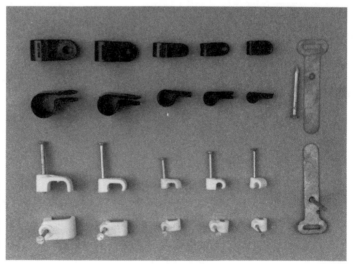

Fig 6.8 *There is a range of different sizes and types of cable clip available, of which the white plastic ones, with a steel nail to hold them in, are to be preferred.*

45

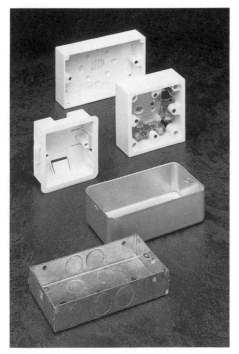

a surface box, the unit is attached to the metal box by a pair of screws, allowing it to lie flush with the wall, the connections being housed within the box.

Junction boxes and terminal blocks

It is essential that electrical cables are joined correctly and safely. **Never twist or solder wires together and cover them with insulating tape. The use of a junction box or terminal block for all permanent wiring is essential**. Junction boxes are available with three or four terminals, for lighting and ring circuits, with current ratings according to purpose. Special four and five way boxes are available for three phase supplies. Flex must only be joined using the necessary two or three way connectors. These may either be permanent connectors, or in the form of a special plug and socket connector. In the latter case, **the female unit must always being connected to the 13 amp plug** and the male to the appliance or lamp.

Wiring up the different circuits

There are three types of circuit required in the average home workshop: the lighting circuit, the ring main circuit, and the radial high power circuit. Each will be fed from a fuseway in the consumer unit, and each will need to be wired using the appropriate cable. The planning of the wiring installation is important, not least to minimise the amount of cable used. Chapter Ten gives further information about lighting circuits, Chapter Eight covers ring main circuits, and Chapter Nine deals with radial high power circuits.

Test equipment and testing circuits

Once all the wiring has been safely installed, **it must be tested before either connecting to power** at the consumer unit, or, if a new consumer

Fig 6.9 *Plastic surface boxes are very simple to install, whilst metal boxes sunk in the wall give a much more professional look.* (Photo courtesy MK Electric.)

wall and the unit is attached to the box by a pair of screws.

Flush mounting is, perhaps, the perfectionist's solution for a workshop, but calls for a metal box to be sunk into the wall. This requires a hole of suitable size and depth to be cut into the wall, and the box to be attached to a solid wall with screws and wall plugs. Special adapters allow metal boxes to be installed in hollow walls. **It is essential that rubber grommets are fitted to each and every cable entry to a metal box to prevent chafing of the wire**. As is the case with

Fig 6.10 *Special junction boxes are available for wiring up ring mains. With three terminals, and a 30 amp rating, they are ideal for wiring in spurs.*

unit is involved, have it tested and connected up by the Electricity Board. **Every alteration and addition must comply with the *IEE Wiring Regulations*, and must not impair the safety of an existing installation**, whether it be a simple addition to an existing circuit, or a new circuit to a consumer unit. Whenever new work has been done, a series of inspections and tests must always be carried out. The model engineer may obtain an application form from the local Electricity Board to connect new wiring up to the mains. This also requires the completion of a test certificate. The list of the things which require inspection is summarised in the following checklist:

- *Are the conductors correctly connected and identified?*
- *Are all cables routed safely?*
- *Are the cables of the correct current carrying capacity?*
- *Are all sockets/fixed connection units and lamp holders correctly connected?*

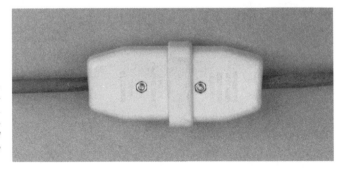

Fig 6.11 *Flex connectors are less than a perfect solution, but are acceptable where flex has to be joined. The female end must be connected to the 13 amp plug.*

47

- *Are there satisfactory methods of protection against direct and indirect contact with live parts?*
- *Are there appropriate devices for switching and isolation?*

Following a satisfactory inspection, the following tests must then be undertaken:

- *Electrical soundness and correct connection of earth leads.*
- *Continuity of all three conductors of every ring circuit*
- *Insulation between phase and neutral conductors and between both phase and neutral conductors and earth. (Test equipment must be capable of providing 1 ma at 500 volts DC, and the resultant insulation resistance must be at least 0.5 megohms.)*
- *Correct operation of any RCD, by simulating an appropriate fault condition. (This must be independent of any test facility incorporated in the RCD.)*

It can be seen that the list is quite a comprehensive one and requires the use of specialist test equipment, the ownership of which is likely to be beyond the needs of the average model engineer. However, for the odd few who may own or have access to such equipment, the following list details the items needed:

- *Insulation tester*
- *RCD tester*
- *Continuity tester*
- *Wiring tester*

It is always to be recommended that any work is checked by an electrician, registered with the National Inspection Council for Electrical Installation Contracting, who can complete the test certificate for minor changes, and pre-inspect prior to connection of a new consumer unit. Such electricians can readily be found through *Yellow Pages* listings. The Electricity Board will carry out their first inspection free of charge, and have no objection to DIY wiring, providing, and only providing, it complies with the Regulations.

Other types of wiring

The three most common other types of wiring likely to be needed in the home workshop are standard four core telephone cable, so called bell wiring, and co-axial cable. The wiring required for three phase circuits is described in Chapter Eleven. Telephone cable is used, as its name suggests, for wiring telephone extensions. Bell wiring is installed for permanent low voltage installations, and co-axial cable connects televisions to their aerials. All three types of cable are readily available from the normal supply sources, and apart from co-axial cable, require no special skills to use.

Co-axial cable consists of an internal single core insulated copper conductor, surrounded by a mesh of copper wire to screen the inner conductor from electrical interference. This outer mesh must be connected to the outer casing of any co-axial plug or socket, and needs to be pulled into the form of a multistrand wire for connection to any permanent aerial socket.

CHAPTER 7

Outside workshops

General

An outside workshop, not attached to the house, poses something of a problem in terms of how to get electricity from the house into the workshop. It can also provide difficulties in terms of communications with the main dwelling, as well as with workshop security. It is likely to require independent heating, and possibly facilities for brewing up refreshments. None of these problems is difficult to solve, and it is assumed that the workshop owner will take the workshop power from the existing household consumer unit.

Size and length of cable

The first questions to be resolved are how far the workshop is from the power supply and what total power is likely to be needed? The answer to the first of these will tell how much cable is going to be needed, and to the second, the size of the cable required. In considering the length of cable, it is important to start the measurement from the point of connection in the house. This may be directly to a junction box adjacent to the consumer unit, or to a radial cable run from the consumer unit to a junction box on the inside wall of the house nearest to the workshop building. A similar allowance will be needed getting the cable into the workshop building, though in this case, there is considerable freedom in the location of the distribution board in the workshop.

The two basic ways of connecting up to the outside workshop are using either an overhead cable, or a buried cable. Both have their advantages and snags. For most home workshops, a cable capable of carrying 30 or 45 amps should more than suffice, and with distances over 5 metres, it is worth fitting the higher rating of cable to avoid voltage reduction when running at maximum load.

Underground connection

For an underground installation, **a mineral insulated, copper sheathed (MICS) or armoured cable is a must**. Both types of cable are PVC sheathed, but the armoured cable is sheathed with steel wire, which is itself covered with insulation. Although it is sometimes permissible to use the copper or armoured sheathing as the earth conductor, the only certain course is to use three conductor cable (two plus earth) as well as the copper or armoured sheathing. Such cables are expensive, and must be terminated with threaded

Fig 7.1 *A metal junction box complete with gland for armoured cable, earth tag and cable clamp for standard PVC sheathed cable.*

glands at purpose-designed junction boxes. Inside the box, a terminal block allows connection to a standard PVC covered cable.

An underground cable demands a trench is dug at least 500 mm deep, and deeper if it is to run, for example, under a vegetable patch. The bottom of the trench must be cleared of sharp stones, and, for preference, should have a bottom layer comprising 50 mm of sand.

With MICS and armoured cable no further protection is necessary, although a duct may be preferred by some. This is probably most simply made by laying a course of bricks along each side of the trench, and then oblong paving slabs over the top. Providing the cable run avoids areas which will be dug over in later years, it offers a very permanent solution, and its length is only limited by the voltage drop along it.

Fig 7.2 *Waterproof glands can be fitted to metal junction boxes, for example at the point where the armoured cable enters the building. A terminal strip allows connection to standard PVC sheathed cable.*

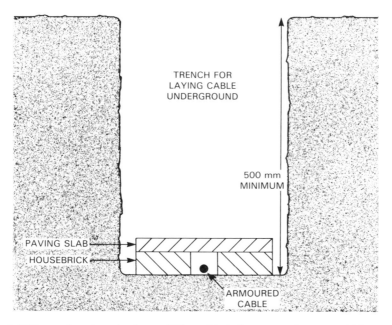

TRENCH FOR
LAYING CABLE
UNDERGROUND

500 mm
MINIMUM

PAVING SLAB

HOUSEBRICK

ARMOURED
CABLE

Fig 7.3 *A 500 mm deep trench, deeper if the cable is to pass under a vegetable patch, with a conduit made from a course of bricks covered with paving slabs, will permanently protect the armoured cable.*

Overhead connection

Overhead wiring uses cheaper cable than those suitable for underground use, **but must be supported once the distance spanned exceeds three metres horizontally**. Up to that distance, PVC cable may be used unsupported, or a 25 mm galvanised steel conduit may be installed; the latter with a PVC bush at each end, and also earthed at each end with a length of 2.5 sq mm green/yellow PVC insulated earth cable. The earth wire should connect to the consumer unit in the house and the switch/fuse box in the workshop. In either case, the cable or conduit must be supported at each end by the building, and **the height rules which apply to catenary systems are also** mandatory. However, even for distances up to three metres, catenary suspension is still to be preferred.

For **greater distances, a catenary system is essential**, with the galvanised steel catenary wire itself earthed, and the cable suspended by clips or slings below the catenary wire. It is ideal for short distances, but above around five metres in length may need intermediate supporting wooden poles. It relies on a route where its vertical location will not cause an obstruction to the movement of anything beneath it. **It must be located at least three metres above any pathway, and 5.2 metres above any drive**. Ordinary PVC insulated and sheathed cable can be used, providing a much

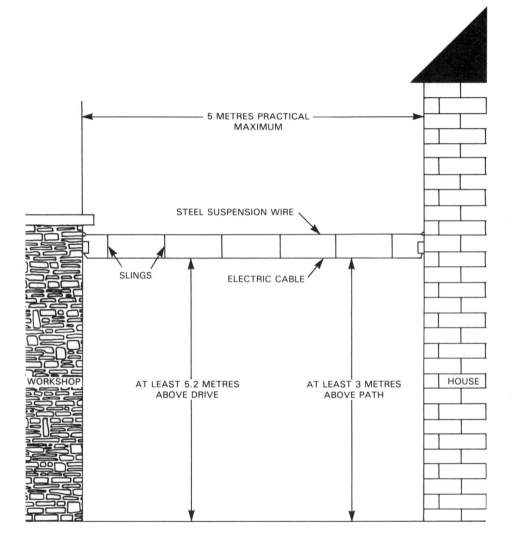

5 METRES PRACTICAL MAXIMUM

STEEL SUSPENSION WIRE

SLINGS

ELECTRIC CABLE

WORKSHOP

AT LEAST 5.2 METRES ABOVE DRIVE

AT LEAST 3 METRES ABOVE PATH

HOUSE

Fig 7.4 *A catenery system comprises a steel wire, firmly anchored to each building, carrying the electrical cable suspended from insulated slings. It provides an economical solution for providing power to outside workshops where the distance to be bridged is not more than about five metres.*

easier and cheaper solution than an underground installation.

Connection to the house

The connection into the house depends somewhat on the amount of electrical power likely to be needed in the workshop. Typically, a 5 amp lighting circuit and a 30 amp ring circuit will suffice, but as a practical maximum, in addition to the lighting and ring circuits, a 20 amp spur for a large item of machinery might be preferred. Thus, a 45 amp connection will need to be made into the consumer unit (assuming that not all electrical machinery will be operated at the same time). If there is no spare way, then a 45 amp switch fuse unit will need to be installed close to the meter, and connected to it by the Electricity Board.

If an underground cable has been laid, then the MICS/armoured cable will have to be connected to conventional PVC cable where it enters both the house and the workshop, using special terminal (or conversion) boxes, which require the use of cable glands for the exterior cable. For the MICS/armoured cable, a compression gland is used, which screws straight into the metal terminal box. It must be tightened to provide a good metal to metal connection to the box, and a waterproof seal around the cable itself. Within the box, a terminal block provides the means of connecting the conductors to standard PVC cable of the same current rating.

Connection in the workshop

It is essential, as with connection to the house, that the cable is run into the workshop in such a way as to ensure that no water can enter the building and its electrical circuits by capillary action up the feeder cable. Termination in the workshop is best done using a small consumer unit or a small switch fuse unit. A typical

Fig 7.5 *A supply unit which can quickly be installed in an outside workshop, suitable for providing a lighting circuit and a single ring main; the latter protected by a built-in RCD.*

switch fuse unit will house a 40 amp RCD and two MCBs; one 6 amp and the other 30 amps. This will power one lighting and one ring main circuit. A four channel consumer unit will provide one circuit for lighting, a second for a ring main, a third for a high current radial for a large machine, and a spare for future expansion. It also provides a convenient means of isolating all the electrical circuits in the workshop when any repairs or extensions are required.

Whether a switch fuse unit or a small consumer unit is selected, the various circuits should be wired in, with connections to the common neutral and earth terminals; the phase wire for each circuit going to the appropriate fuse or MCB. When this has been completed, the incoming cable can be wired to the three terminals; neutral, earth and the

53

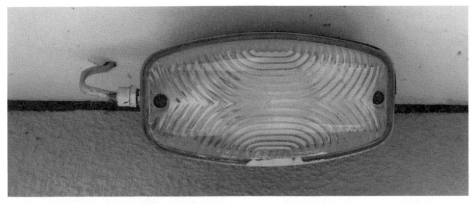

Fig 7.6 *An outside light can be protected from the worst of the weather by locating it under the eaves of a building. However, a watertight exterior quality light fitting is essential.*

switched phase. Once the electricity has been run into the workshop, the standard rules which apply throughout this book once again prevail.

Outside lights and sockets

Fittings that are installed outside the workshop require special protection from the effects of the elements, be they ordinary lights, security lights or power sockets. The two key factors are that the device itself should be weatherproof, and that action should be taken to ensure that water does not run back into the interior electric wiring. In the case of lights, they can often be mounted under the eaves of the building, providing protection from the worst of the weather, whilst power sockets can be protected by a simple wooden box on the wall, with a door giving access. Nevertheless, **only fittings designed for outdoor use may be employed in such locations,** and the manufacturer's instructions must be followed to the letter. 13 amp sockets, with splash proof covers, are a more economic solution than fully weatherproof ones, though, obviously, the latter

must be used in exposed positions. **Any outside socket must be fed from an RCD, and the same rule applies to any piece of workshop electrical equipment being used in the open air.** The most common

Fig 7.7 *External 13 amp sockets, such as this splash-proof one, will need some protection from the worst ravages of the weather.*

54

items likely to be thus used are electric hand drills, electric welders, and air compressors. **Electrical equipment must never be used outside, except in dry weather conditions.**

Security lights

Security lights, which either turn on at dusk or some pre-set time, or alternatively are actuated by anyone approaching, are becoming increasingly popular. They are simple to fit, and provide an excellent deterrent to burglars. Their installation is similar electrically to any other light fitting. They can be mounted outside a workshop, preferably under the eaves to give a modicum of protection from the elements.

Fig 7.8 *A security light, with an automatic sensor to switch it on, is an affordable addition to any outside workshop.*

CHAPTER 8

Plugs and sockets

General

All workshop equipment, apart from those items requiring more than 13 amps (approximately 3 kilowatts), and any permanently installed lighting, will be powered by a ring main circuit. These circuits, as their name suggests, run in a circle from the consumer unit and back to it. The basic circuit is powered from a 30 amp fuse or MCB in the consumer unit. It may be fitted with an indefinite number of sockets and fixed connection units, **provided only that the floor area served does not exceed one hundred square metres**. Where this area is exceeded, a second or third ring circuit must be used. Permanently installed workshop tools and heaters should be powered from fixed connection units, and further details of these are given in Chapter Nine.

It is recommended that as many sockets as possible are provided in the home workshop, as, with the passage of time, there always seems to be a shortage. For the same reason, it is suggested that twin sockets are employed, as these are less expensive than two single ones, as well as taking less time to install. **Lighting circuits or permanently installed lights should never be connected to a ring main circuit**, but rather powered from a separate lighting circuit. However, it is normal practice to power portable lamps from a socket, using a 13 amp plug, fitted with a 2 or 3 amp fuse.

What to power from 13 amp sockets

There are many tools in the home workshop which can be powered from a 13 amp socket. This excludes all but the smallest lathes, milling machines and drill presses, and even these are better permanently wired in, unless they are regularly moved about. The list does include all portable equipment and the following identifies most of the items, drawing less than 13 amps, likely to be found in a home workshop. Some have more to do with tidying up and creature comforts than with model engineering!

- *Bench grinders, polishers and sanders.*
- *Small bandsaws and power hacksaws.*
- *Electric welders and small air compressors.*
- *Lead acid and nicad battery chargers.*
- *Portable powered tools: drills, screwdrivers, saws, sanders and planers.*
- *Soldering irons and electrical test equipment.*
- *Vacuum cleaners and 'dustbusters'.*
- *Portable electric heaters and electric kettles.*
- *Radios and TVs.*

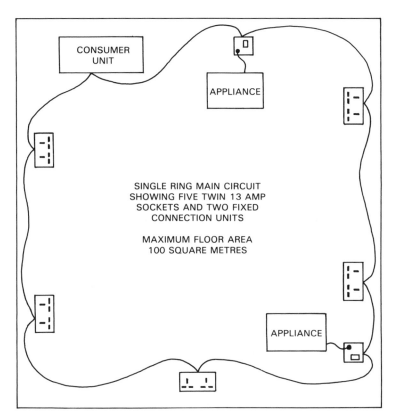

Fig 8.1 *Ring main circuits require two cables to each socket or fixed connection unit in order to provide a ring connection to and from the consumer unit.*

Note that appliances should be switched off at the socket and normally be left unplugged when not in use.

Spurs

Spurs may be attached to any ring circuit, allowing additional sockets to be added simply at a later date, and also allowing economy of cable when installing very remote sockets. The three rules which apply for spurs are as follows, and **adherence to them is mandatory:**

- *The cable used to connect spurs must be at least the same size as that used for the ring main itself.*
- *Only one single or double socket may be fed from each spur.*
- *The total number of spurs must never exceed the total number of sockets fed by the ring main.*

13 amp sockets

There is a wide variety of 13 amp sockets available on the market today. They can be single or double, surface mounted or

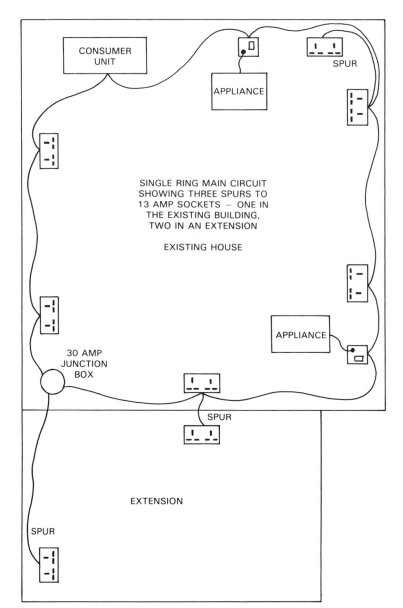

CONSUMER UNIT

APPLIANCE

SPUR

SINGLE RING MAIN CIRCUIT
SHOWING THREE SPURS TO
13 AMP SOCKETS – ONE IN
THE EXISTING BUILDING,
TWO IN AN EXTENSION

EXISTING HOUSE

APPLIANCE

30 AMP
JUNCTION
BOX

SPUR

SPUR

EXTENSION

SPUR

Fig 8.2 *Spurs require only a single cable from a socket or junction box. However, the number of spurs may never exceed the number of sockets fitted to a ring main circuit.*

flush, with or without integral switches, waterproof, splash-proof or for indoor use only, with or without RCDs and in a range of plastic and metal finishes. This bewildering variety can be quickly narrowed down for home workshop use. The preferred choice is for double, flush, switched, plastic, without RCDs and for indoor use. Two exceptions to this selection are first the use of surface mounted sockets where it is difficult or inconvenient to sink them into the wall and second, the use of an RCD switch where none is fitted in the consumer unit, and where there is any risk of an electric shock. The use of such sockets is especially recommended for powering electrical tools with metal cases, particularly those without double insulation.

Sadly, some houses and workshops still use round pin 5 amp and 15 amp plugs and sockets. **These are so obsolete as to mean that the complete house needs rewiring**, and work with such plugs and sockets is considered to be outside the scope of this book.

Installing 13 amp sockets

Before any installation work can be undertaken, the position of each socket must be planned. **They must be located at least 150 mm above floor or workbench level**, or conveniently some 100 mm beneath workbench level. Sockets should be placed on all four walls of the workshop to obviate the need for trailing flexes across the workshop. Sockets should also be positioned conveniently for the location where powered hand tools are going to be used. Plenty of thought is necessary, and for a well equipped workshop, planning should be in terms of at least ten double sockets.

The next decision is whether to fit flush sockets or use surface mounted ones. Where the building is a wooden

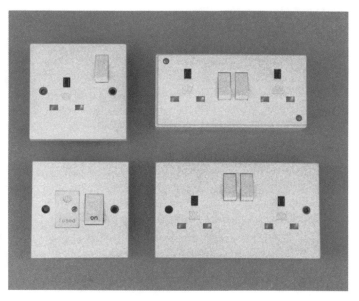

Fig 8.3 *A selection of switched single and twin 13 amp sockets, with a fixed connection unit shown bottom left.*

one, there is usually little choice but to go for surface mounted sockets. With masonry or plasterboard walls, flush mounted units will give a much tidier finish, and take up less space. To flush mount on a masonry wall, a rectangular hole must be cut to take a metal box, sunk into the wall to the depth of the box. This can be done with a cold chisel, with the assistance of holes drilled at regular intervals using a masonry drill, employing a similar technique to that used to make a rectangular hole in steel sheet. The box is attached to the wall using steel screws and wall plugs.

The cables are fed into the box through grommets placed in the cable entry holes, which can be knocked out in the appropriate places; a fair choice of positions being provided. The sockets, apart from spurs, will have two cables to allow the ring to be completed. A loop of cable, sticking out of the metal box, which can later be cut in the middle, will provide the two wires. These should then have their outer insulation cut back, and the inner individual conductor insulation removed. This allows each pair of wires to feed to the appropriate terminal in the socket; **RED to PHASE, BLACK to NEUTRAL and the bare EARTH wire, sleeved with a short length of GREEN/YELLOW insulation, to EARTH.** The socket is then attached to its box by the two screws in its sides, ensuring the spare cable is neatly tucked in behind. A virtually identical procedure is used when a spur socket is installed, but only a single cable is needed.

A rather different technique is required for attachment to a stud and plasterboard wall. In this case, mark out the position of the socket ensuring it does not foul the timber studding. Cut out the hole in the plasterboard, and then fit dry wall fixing flanges to the metal box. Manoeuvre it through the hole so that the flanges lie behind the plasterboard. It is the final attachment of the socket itself that holds the box firmly in place.

Finally, and it's ideal both for plaster-

Fig 8.4 *When installing a socket, the three cables must be fitted into the correct holes and the retaining screws firmly tightened. Note the short length of yellow/green insulation which has been slipped over the earth wire.*

60

board and for wooden walls, use a surface mounting box, and accept that it will not be flush. The plastic mounting box is attached to the wall with screws and, in the case of plasterboard, the correct type of wall plugs must be used.

13 amp plugs

The UK standard is a fused plug with three rectangular pins for phase, neutral and earth. Fuses from 2 to 13 amps may be fitted, with 3 or 13 amps the norm. Differences in the design of the terminals for connecting the flex to the plug, as well as in the cable retention clip are common. One particularly useful variant for the use in the home workshop is the rubber-bodied plug, which is far less susceptible to mechanical damage when dropped. It is thus very well suited to use on portable power tools. Similar rubber bodied sockets are available for use when making up extension leads.

Wiring a plug

Wiring up a plug to a tool or another piece of electrical equipment is a really straightforward task and should not need describing to the average model engineer. However, in light of the large incidence of incorrectly wired plugs found in the UK, the procedure will be described, and is as follows:

- *Cut back the outer sleeving on the flex for a distance of 4 centimetres.*
- *Cut back insulation on the three wires for a distance of 1 centimetre. Note, some double insulated tools only have two wires, phase and neutral.*
- *Twist the ends of each bare wire to form a solid conductor.*
- *Remove the cover of the plug.*
- *Insert each wire into the hole in the terminal, or bend it clockwise under the terminal nut.*
- *Ensure **BROWN goes to PHASE, BLUE to NEUTRAL and YELLOW/GREEN to EARTH**.*

Fig 8.5 *Having stripped back the insulation, bend each wire around its terminal post and replace the nut. There is some variation in the design of terminal from plug to plug.*

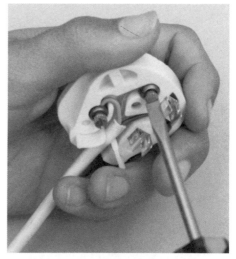

Fig 8.6 *Tighten each nut securely, and ensure that the cable sits tidily in the cable anchor. Some anchors require two screws tightening to retain the cable.*

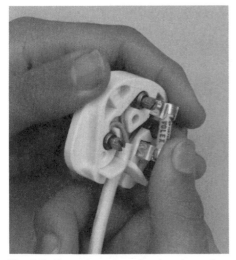

Fig 8.7 *Fit a fuse into its locating clips, ensuring it is of the correct rating.*

Fig 8.8 *Finally, replace the cover and tighten up the retaining screw.*

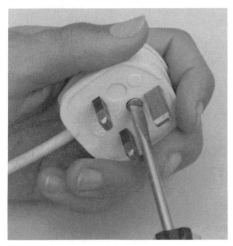

Note that many plugs still carry an 'L' at the phase wire terminal. A quick way of remembering which wires go to which terminals is **BR**own; **B**ottom **R**ight, **BL**ue; **B**ottom **L**eft.
● Firmly tighten the terminal screws/nuts.
● Ensure **the cable is secure in the cable anchoring device**.
● Insert **the correct value of fuse**.
● Replace the cover and secure.

The cartridge fuse, installed in each plug, must be rated according to the device being powered by the plug. Regrettably, new plugs usually include a 13 amp fuse, which should be replaced by one of a lower rating for all but the most power consuming devices. The correct fuses are:

Current	Colour	Use
2 amp	Black	Portable lights, radios and TVs
3 amp	Red	Portable lights, radios and TVs, low power workshop tools and appliances up to 720 watts capacity
5 amp	Black	Small tools of up to one kilowatt consumption
10 amp	Black	Medium powered appliances up to two kilowatts
13 amp	Brown	Most motor-driven equipment and electric heaters

2, 5 and 10 amp fuses are less commonly used than 3 and 13 amp ones.

The flex connected to the plug should be carefully selected to match the particular use to which it is being put; simple twin or three core for portable lamps, more substantial three core cable for tools, and special armoured cable where mechanical damage is a risk. Tools which may be used out of doors should use the bright orange flex designed to be highly visible in the outdoor environment.

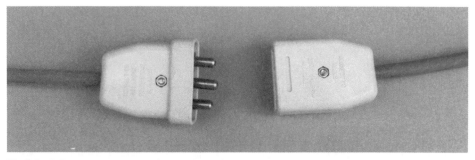

Fig 8.9 *A flex connector may be used as a last resort to join two lengths of flex. Two or three pin variants are available, depending on application. The wiring of the female end to the 13 amp plug is a must to avoid the considerable danger of exposed live pins when the connector is pulled apart.*

Flex connectors and extension leads

Flex connectors may be used to join two lengths of cable together, but **always with the female half of the connector wired to the plug**. For preference, how-ever, the existing cable should be replaced with a new one of sufficient length to avoid the need for a joint. As mentioned in the chapter on safety, **twisted wires and insulating tape are**

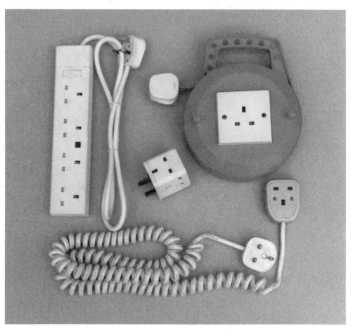

Fig 8.10 *A variety of extension leads and adapters of the type likely to be found in many homes. Care must be taken in their use to avoid the potential dangers described in Chapter Three.*

quite unacceptable for joining cables. Extension leads come in a wide selection of lengths and power ratings, from simple twisted spring coiled ones giving around three metres of extension to 100 metre ones supplied on wind up reels. **It is important, for safety reasons, to uncoil completely any extension lead wound on a reel before connecting it to avoid over-heating**. All more recently manufactured ones should carry such a warning. The use of extension leads in the home work-shop on a permanent basis is to be deprecated. Sufficient permanent sockets should be installed, as the dangers of an extension lead trailing around the work-shop is obvious. Their use to reach awkward places, when doing electrical installation work in the home workshop is another matter. They also represent an easy way of providing electricity for a power tool, when the supply to the workshop itself has been isolated.

Adapters

Multi way adapters, used judiciously, have their advantages, and enable several devices to be run from a single socket. **Under no circumstances must they be used in such a way as to overload a socket beyond its nominal rating of 13 amps**, and all modern ones carry their own fuses, identical to those fitted to 13 amp plugs. The use of adapters is again an indication of the requirement for more sockets to be in-stalled in the workshop. However, running a pair of anglepoise lamps via an adapter from a single socket is an example of a sensible use of an adapter.

Time switches

The use of time switches in a home workshop is likely to be limited. They can be convenient for turning on an electric heater, sometime before the workshop is

Fig 8.11 *A time switch has a number of applications in the workshop, and the initial expense can quickly be repaid, both in con-venience and the saving in power consumption.* (Photo courtesy MK Electric.)

to be used in winter, to warm it up. They can also be used for switching on portable lights and a radio to deter thieves. Time switches are available as self-contained plug-in units, similar to an adapter with a clock face, and normally operate on a daily or weekly cycle. Those with digital displays are usually more expensive than those with analogue ones. A useful variant incorporates a thermostat, turning on the power when the temperature falls to a pre-set level.

Speed controllers

Speed controllers can be very beneficial when used with machine tools, Sadly, the majority of those readily available on the market, and all those available at a low cost, will not work with induction motors. In fact, trying to use such controllers with induction motors will stop them running and quickly cause damage through overheating. However, their use with portable powered tools, such as electric drills, can be very advantageous. A simple rotating knob allows the speed to be varied electronically from full power to almost stopped, with less loss of torque at low speed settings than would normally be expected.

CHAPTER 9

Fixed appliances

Fixed connection units

Permanently installed machine tools and other appliances should not be connected up using a 13 amp plug and socket, but rather wired into a fixed connection unit. These come in a variety of designs, switched or unswitched, with or without an indicating neon, and with a cable outlet through the front of the unit, or requiring wiring from behind. In all other respects, their connection and use is identical to that of 13 amp sockets, and this includes a maximum current rating of 13 amps, as well as the need to select the right value of fuse to match the piece of equipment being powered. When working, in the electrical sense, on an appliance, **it is essential for safety reasons that the power is first turned off and the fuse removed.**

Only one piece of equipment may be connected to each unit, and the correct rating of fuse must be fitted. Unlike sockets, double fixed connection units are not available. It is suggested that switched units are always employed, so that the item of equipment can easily be isolated from the mains, and that for preference, those with indicating neons are selected. The fixed connection unit should be located as close to the item of

Fig 9.1 *A fixed connection unit should always be used for powering permanently installed items of workshop equipment which draw less than 13 amps. The removable panel beside the switch allows the correct value of fuse to be fitted.*

equipment it is powering as is feasible, and the connecting cable should naturally be of the correct rating.

Equipment needing more than 13 amps (3.12 kW)

Some of the larger items of workshop

equipment need more than 13 amps (3.12kW) and a separate radial circuit from the consumer unit will have to be provided for each of these. Normal circuit ratings are 15, 20, 30 or 45 amps, allowing loadings of up to 3.6kW, 4.8kW, 7.2kW and 10.8kW respectively. For each circuit, a separate 15, 20, 30 or 45 amp switch will be needed, and these are available with or without neon indicators, although the former is much to be preferred. Electric cooker switches are a readily available alternative for the higher current ratings.

These circuits require an MCB or fuse of the relevant capacity to be fitted in the consumer unit when the circuit is

Fig 9.3 *Many lathes are supplied with a custom-built three position switch allowing direction of rotation to be selected.*

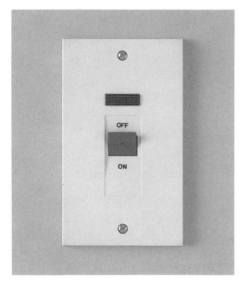

Fig 9.2 *Cooker switches are available, which will handle 30 and 45 amps, and are suitable for powering larger items of workshop equipment. A neon indicates when power is being supplied to the machine.* (Photo courtesy MK Electric.)

established, together with cable of 4 sq mm cross sectional area for 30 amp circuits or 6 sq mm for 45 amp ones. The circuits are always radial ones and will normally only power a single item of equipment. **The switching unit must be fitted within two metres of the workshop appliance** and the cable from the switch to the appliance must be of the same size as that used in the radial circuit.

It is considered most unlikely that any machine tool needing more than 10.8kW will be installed in a home workshop. The switches and control gear required for high power machine tools are usually supplied with the machine itself, but if purchased separately, should be of sufficient power rating, and of the type that automatically turn off in the case of mains failure.

Types of fixed appliance

Typical fixed appliances which require wiring in permanently include:

- *Lathes, milling machines and shapers*
- *Large bandsaws and powered hacksaws*
- *Drill presses*
- *Extraction fans*
- *Electric heaters*

In fact any item of electrically powered equipment which is to be permanently installed in position in the workshop should be powered from a fixed connection unit, rather than from a 13 amp socket, providing both a cheaper and safer solution.

Heaters and thermostats

As has already been mentioned, the great advantages of electrical heating are the ease with which it can be installed, the fact that it does not emit any water vapour during the heating process and the ability to turn it on and off with a thermostat and/or time clock. Against this is the high cost of running this form of heating. However, in a workshop where usage is limited to the evenings and weekends, electrical heating can still be an attractive option, and electric radiators can be wall mounted, as can infra red fires. Oil-filled electric radiators give out good heat, comparable with central heating radiators, but require time to warm up. Fan heaters can also be used to advantage, due to their ability to quickly to heat up a workshop; this speed being their greatest asset.

The inclusion of a suitably located thermostat in the circuit will turn off the heater automatically once the workshop has reached the required temperature, and turn it on again when the temperature

Fig 9.4 *A fan heater is ideal for quickly warming up a cold workshop, although it is expensive to run for any prolonged period of time.*

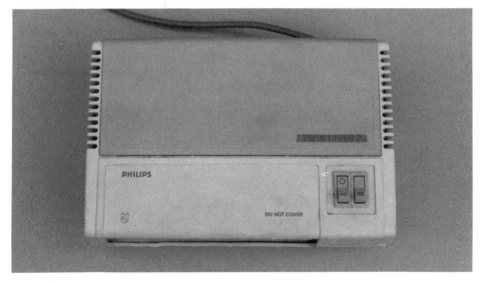

drops. Such devices are readily available, and quickly save their cost in terms of reduced electricity consumption. They are connected to the device being controlled, and the manufacturer's fitting instructions should always be followed. When a workshop is used at the same time every day, a time clock can be fitted to turn on the heating at a fixed interval before work commences and to turn off a set time later. This can be of great advantage to the model engineer of fixed habits and ensures the workshop has warmed up before work commences; a boon in winter.

A number of wall and window mounted extractor fans are available which operate when the relative humidity in the workshop exceeds a given figure, usually around 65%. They include a built-in

humidity sensor and are ideal for helping to avoid damaging condensation on machine tools.

Night storage heaters

Connecting night storage heaters is somewhat different to other heaters, as they are run from a separate metering system supplied by the Electricity Board. This low tariff electricity is normally distributed by a separate switch fuse unit, with a single radial circuit to each individual storage heater. The power is only available at night, for some seven hours, warming up the special heat retaining core within the heaters. Units which just release their heat by day, as well as fan assisted units, are both available, and typical power consumption is 4.8 kW, requiring a 20 amp fuse in the consumer unit and 2.5 mm cable connection. They are fairly bulky devices and thought needs to be given to their location.

It is considered that night storage heaters would only be chosen for use in a workshop if similar units are already installed in the house. For this reason, it is assumed that the low tariff switch fuse unit is already installed and suitable additional fuseways are available. Thus, connection of a new unit in the workshop involves running a single 20 amp radial circuit from the low tariff consumer unit directly to a fixed connection unit adjacent to the chosen location for the heater, using a PVC cable with 2.5mm conductors. Should a night storage heater be installed in an outside workshop, a separate cable will have to be run from the house to the workshop to power the heater.

Fig 9.5 *The use of a thermostat to control any form of electrical heater used in the home workshop will soon pay for itself, and, over a long period, save a considerable sum of money.*

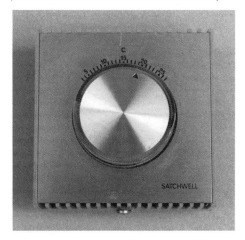

CHAPTER 10

Lighting

General

Good lighting in a workshop is a critical factor in achieving a high standard of workmanship. With increasing age, it is a sad fact of life that people's eyes become less efficient, and thus the importance of good lighting increases. Adequate lighting is necessary from a safety point of view as well, and helps to reduce the fatigue of concentrating on delicate tasks. Well-planned lighting can also help to eliminate those awkward dark corners where something that has been dropped is so difficult to find.

Workshops need two types of lighting. The first is background lighting, which will always be turned on when there is anyone in the workshop. Fluorescent lighting is particularly well suited to this role. The other type is task lighting, which provides added illumination when working on a specific machine or in a particular part of the workshop. Spotlights or portable lights are usually best for this role. Information on low voltage lights, which can be particularly useful for mounting on machine tools, is covered in Chapter Twelve.

In houses built before 1966, each light is connected into a lighting circuit via a junction box. Since that date, loop-in roses, which combine the functions of a ceiling rose and a junction box, have provided a more economical solution to the same problem. It is worth noting that in lighting circuits, each light is not individually fused, but relies on the circuit fuse in the consumer unit for protection.

Lighting circuits

A workshop's lighting can usually be fed from an existing lighting circuit without any difficulty and Fig. 4.3 shows a typical household lighting installation. If an outside workshop is to be lit, then a lighting circuit can be connected to the distribution board within the workshop. Lighting circuits are radial ones, leading from the consumer unit though each room, with their junction boxes or loop-in roses, until they reach the final room. A house will normally have two lighting circuits; one upstairs and one downstairs. **Each lighting circuit is limited to an absolute maximum of twelve lamps and, in addition, to 5 amps or 1200 watts.** Thus, if bulbs of more than 100 watts are used, the number of lamps being fed from a single circuit must be cut pro-rata. The recommended cable is PVC sheathed, with two separate PVC insulated conductors, each 1 sq mm in

Fig 10.1 *Pendant light fittings and straight batten holders are those most commonly found in rooms being converted to workshops.*

natively, a junction box can be inserted in an existing lighting cable. If the lighting circuit already has its full quota of light fittings, or if there is no available lighting circuit, then a new circuit will have to be established, fed through a 5 amp fuse or MCB in the consumer unit.

Junction boxes and their wiring

Wiring up a light fitting is not quite as simple as at first might be imagined. A junction box is needed for each light switch and its associated light fittings to connect them to the mains. Modern loop-in ceiling roses eliminate the need for a junction box by incorporating this function within the rose. Older roses still retain the need for a junction box to be used as well. Clear thinking is required when connecting up each light, and following the diagrams to the letter is most important.

Having decided where the light fitting and light switch are to be located, the type of fitting to be installed must be selected. Common types include the pendant, the batten lamp holder and the fluorescent tube fitting. Mark the position of the light fitting and the switch. Measure the amount of cable needed, remembering that wiring will not normally cut corners and that a little extra length should be allowed, as a centimetre too short will mean cutting a fresh piece of cable.

It is best to position the light fitting so that it can be attached to a joist, while allowing the cable to go straight through the ceiling plaster beside the joist. Make a hole in the ceiling and push through a length of cable to go to the junction box, which is also wired to the switch and the mains. If a loop-in rose is used, then two cables should be passed through the hole for connection to the mains and to the switch. It is well worthwhile labelling the

area, together with an earth lead.

When lighting is to be installed in a workshop from scratch, or when another light is added in an existing room, the first thing to do is to establish where the electrical source is located. **Do not connect lighting directly into a ring main**. The connection to the mains must be at a junction box or loop-in rose, where there is already a lighting circuit. Alter-

Fig 10.2 *The use of a junction box is necessary for most workshop light fittings. Care is needed as it is essential to connect the wires correctly.*

Fig 10.4 *Replacement of the flex in a ceiling rose is a simple task, the wires requiring connection to two screw terminals at each end, and routing through the anchoring devices.*

Fig 10.3 *The newer type of loop-in ceiling rose includes the functions of a junction box within the rose itself, saving both time and money during installation.*

72

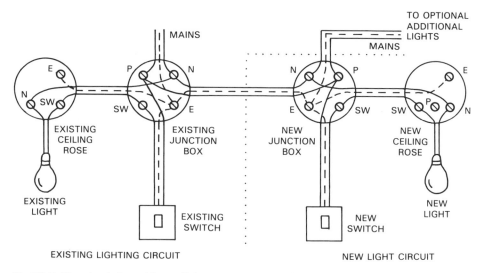

Fig 10.5 *The circuit for adding a light to an existing lighting circuit using a junction box.*

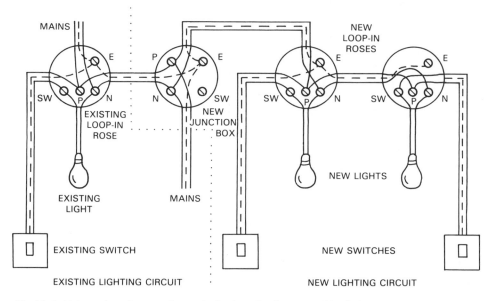

Fig 10.6 *Using a loop-in rose, instead of a junction box, to add a light to an existing circuit.*

cables 'mains' and 'switch' (and 'light' if a junction box is used) with a ballpoint pen or indelible marker. Finally, the cable from the junction box or ceiling rose is wired up to the light fitting. Fittings with an open ceiling plate require a small circular plastic box fixed flush in the ceiling, installed to house the connecting wires and terminal block. Somewhat different rules apply to the installation of down lighters, and these are covered later.

Light switches

The basic on/off plate switches are familiar to all, and come in single and multiple units. They are simply connected to the lighting junction box by a twin core and earth cable. Where space is very restricted, a narrow architrave switch can be used. Switches, which include a facility to turn on automatically at dusk and off at a chosen time, are a good idea from a security point of view, as are programmable time switches.

Dimmers are also available which are interchangeable with the basic on/off switches, but there is usually no practical reason to use them in a workshop. Neither dimmers, nor electronic time switches will normally operate fluorescent lighting.

Plate switches may either be fitted on using a surface box or flush mounted in a buried metal box. These boxes come in two depths; 16 or 25 mm, depending on the depth of the switch. An alternative to plate switches is to employ a ceiling mounted pull cord-switch. These have been designed for use in places where wet hands are found; kitchens and

Fig 10.7 *Plate switches come in a wide range of configurations; that at bottom right automatically turns on the lights at dusk for a preset length of time to deter burglars.*

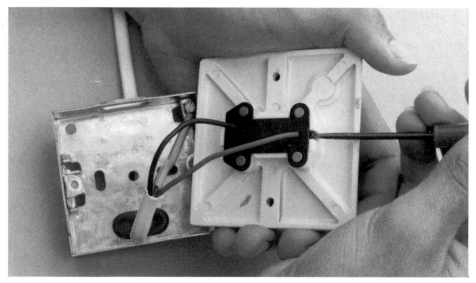

Fig 10.8 *The insulated phase and switch wires are connected to the switch itself, and the earth to the metal box.*

Fig 10.9 *Pull switches provide a useful alternative to wall mounted units, and can be quicker and cheaper to install.*

bathrooms. They may, nevertheless, be useful in a workshop, since their installation does not require a cable to be run down the wall and they can be located close to where the light is needed.

Two way switching may be required if there are two entrances into a workshop, either of which may be used. In this case a light switch should be fitted at each entrance, and connected up to form a circuit which allows the lights to be turned on and off from either switch. This does require non standard three wire plus earth cable and, of course, two way switches. That apart, the wiring does not cause any particular difficulties.

Fluorescent lighting in the workshop

Fluorescent lighting is ideal for providing general background lighting in workshops. The three main advantages of fluorescent tubes are that they do not cast hard

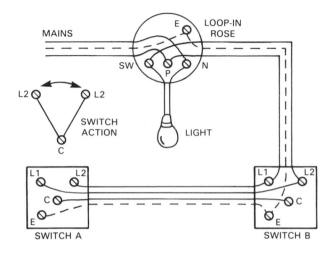

MAINS

E

LOOP-IN
ROSE

SW

P

N

L2

L2

SWITCH
ACTION

C

LIGHT

L1

L2

C

E

SWITCH A

L1

L2

C

E

SWITCH B

Fig 10.10 *Wiring up a pair of switches for two way switching requires the use of cable with three insulated cores plus earth. The extra wire normally has yellow insulation.*

shadows, they are significantly cheaper to run than incandescent lighting and they give out very little heat. They give roughly twice as much light, for a given wattage rating, as incandescent bulbs. On the negative side, the initial cost is high. As a basic rule of thumb, a workshop will require at least ten watts of this type of light per square metre of floorspace. Standard fittings can be purchased in a variety of lengths and associated wattages, normally with one or two tubes per fitting.

Whilst incandescent lights work by passing a current through a high resistance length of tungsten wire until it glows white hot, the principle of operation of a fluorescent light is entirely different. The tube is coated on the inside with a fluorescing powder and then filled with a suitable gas. The choice of materials affects the final colour of the light emitted. Electrodes at each end of the tube enable a stream of electrons to flow through the tube, causing the coating to glow, thus providing the light. Unlike an

Fig 10.11 *A 1.5 metre fluorescent tube gives adequate background lighting for a workshop with a floor area of up to six square metres.* (Photo courtesy Osram Ltd.)

incandescent lamp, a fluorescent tube will not light up automatically when connected to the mains, unless special starting equipment is provided, and this is generally built into the fitting. If a fluorescent light should start to flicker, it should never be left to flicker, as this can damage the control equipment in the fitting. It is also better to leave a fluorescent light on than to turn it on and off frequently, as each time it is turned on the tube life is reduced.

There are two systems currently in use; the switch starter type and the quick start type. The first incorporates a plug-in starter switch and choke to help the tube to strike and start working. The quick start uses an earthing strip and transformer in place of the starter and choke. It really doesn't matter which type of fitting is used in the home workshop. The quick start, as its name implies, lights up without the characteristic delay of the starter type but at the expense of a slightly shorter tube life. It also will not work without an earth connection, which may not be available in older lighting circuits, but which can easily be added. Its use is far less common than the starter type.

Normally sold in lengths of 1.2, 1.5, 1.8 and 2.4 metres, the shorter length fluorescent tubes are probably best suited to home workshops. They can be fitted in place of existing pendant lights, by removing the ceiling rose, screwing the fluorescent fitting in its place, and then connecting the new fitting in the normal way, i.e. connecting the existing wires that ran to the ceiling rose into the terminal block in the tube holder. Should the wires coming out of the ceiling be insufficiently long, then it will be necessary to fit new wiring back to the relevant junction box. If the ceiling rose was a loop-in one, it will have to be replaced with a junction box. Shorter tubes, down to 450 millimetres in length, as well as circular ones are available, but are less well suited to workshop use.

Unlike incandescent lights, which either work or don't, fluorescent lights can develop repairable faults. Thus, fault finding is important if these types of light are fitted. There are five main fault modes, and some of them can easily be repaired:

Fig 10.12 *The inside of a fluorescent tube fitting showing the main components used for starting the tube.*

REPLACEABLE STARTER

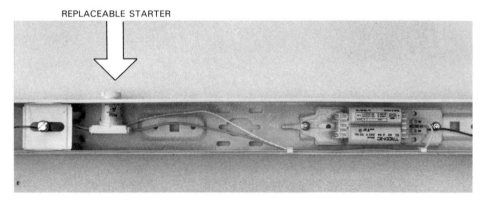

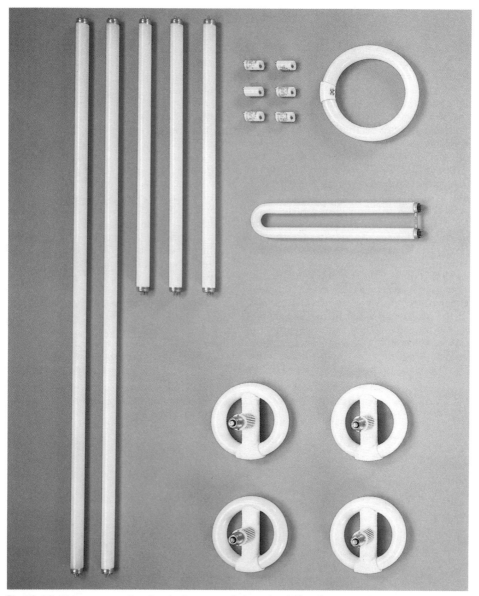

Fig 10.13 *Fluorescent tubes for the home workshop should be chosen to give the required level of illumination, and also to fit in the available space.* (Photo courtesy Osram Ltd.)

1. The tube glows at each end, but makes no attempt to illuminate along its full length.

   Switch starter type: If the glow is white, the starter needs replacing. If red replace the tube.

   Quick start type: Check for a faulty earth connection.

2. The tube glows at one end and keeps trying to start.

   Switch starter type: Short circuit at dead end of the tube. Normally requires tube replacement.

   Quick start type: Open circuit at the dead end of the tube. Check wiring and if no fault found, replace tube.

3. The tube flickers between off and full brightness.

   Either a faulty starter, or the tube needs replacing. This fault can also be caused by low mains voltage.

4. The tube lights up but at only about 50% illumination.

   Replace the tube.

5. The tube appears to be completely dead.

   May be caused by a blown fuse or wiring discontinuity, a fault in the light fitting or a broken electrode within the tube.

Compact fluorescent lamps

The ones recommended for use in the workshop are those which are direct plug-in replacements for incandescent light bulbs. Due to their small size, they tend to give shadows, but do retain all the other advantages of fluorescent tubes. Again their initial price is high. A number of manufacturers offer compact fluorescent lamps which require special light fittings. There is no good reason for choosing these for lighting in a workshop. However, should they be used for some reason, then the special fittings, or adapters to allow them to be used with standard light sockets, must be purchased at the same time.

The range of light bulbs

Light bulbs come in all shapes and sizes, as well as wattage ratings; the last being proportional to the amount of illumination given for a particular type of bulb. It is important to remember that only about a quarter of the electricity consumed by an incandescent bulb is converted into light, the remainder being dissipated as heat. Some bulbs are obviously more efficient than others, and the relatively recent development of fluorescent bulbs which fit into conventional light sockets has not only significantly improved efficiency, but also brought about an approximate eight-fold increase in bulb life. As with anything in life, these benefits can only be acquired for a significant increase in purchase price – by a factor of some ten times! There is still, however the benefit of reduced running costs.

Bulbs, and for that matter light fittings, come with one of two types of connection. The UK standard is the BC or bayonet connection, which allows a bulb to be inserted with a push and quarter turn movement. The other ES type uses a screw-in action to make the connection, and is increasingly common in specialist lamps, such as spotlight and down-lighters. Generally, 60, 75 or 100 watt bulbs should be used in the workshop, except in miniature spotlights, where 40 watts is the normal limit. Bulbs of 15 and 25 watts at the lower end, and 150 watts at the higher end can also be purchased.

Assuming that aesthetics are not a key factor in selecting lighting for a workshop, the most useful incandescent bulbs are the conventional pear shaped, either clear or pearl, and the internally

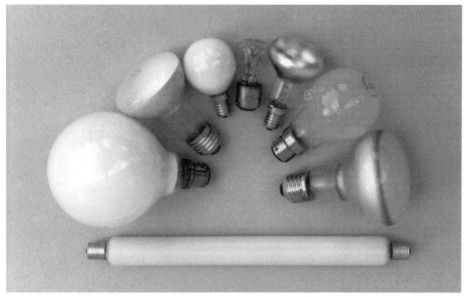

Fig 10.14 *A selection of the many light bulbs available, showing the wide variety of size and shape.*

silvered reflector bulbs. This does not mean that other bulbs, such as the mushroom, floodlight, linear striplight and pigmy may not be used for particular applications.

On the fluorescent side, an ever increasing selection of compact lamps is made by the main manufacturers, which sell under a wide range of trade names including Dulux EL, EL Globe and Circolux EL lamps by Osram and SL lamps by Philips. These may be plugged directly into a standard light fitting. Others, such as the 2D by GE THORN require special fittings or adapters, and are thus probably not the best choice for the home workshop.

Incandescent light fittings, fixtures and lamp types

Many and varied is the type of light fitting available today. Starting with the ubiquitous ceiling pendant and straight batten holder, these are usually found already installed in any household room or garage which may be selected for use as a workshop. They are far from ideal for our purposes. Even less ideal are wall lights because they take up valuable space needed for storage. A wide choice of spotlights, downlighter, anglepoise and other portable lights are also likely to find applications in the home workshop.

Task lighting

Task lighting is needed to illuminate a selected work area or machine tool. Spotlights, track spotlights and downlighters are all good at providing concentrated pools of light, but if situated behind the modeller, cast sharp and awkward shadows. Track spotlights are fitted to

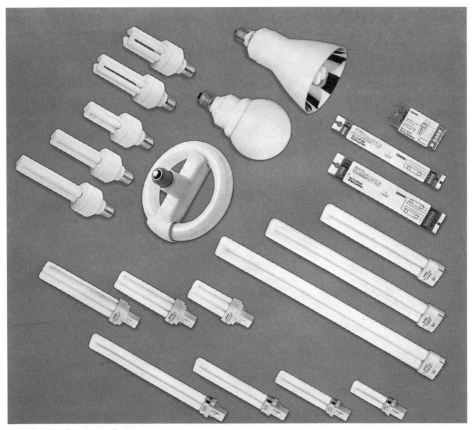

Fig 10.15 *Compact fluorescent tubes are on the market in an increasing range of shapes and sizes. Many will plug straight into a standard light fitting.* (Photo courtesy Osram Ltd.)

aluminium tracks, normally supplied in one metre lengths, which carry electrical power along their length. Thus, the lights can be moved along the track, and their position varied to suit the needs of the workshop. One end of the track has a terminal block, which is wired up as any other light fitting. However, care must be taken to **ensure the number of spotlights fitted to the track does not cause the total number in the complete lighting circuit to exceed twelve**. Tracks may also be wall mounted and connected to a lighting circuit.

Downlighters can be used for task lighting. Their installation is identical electrically to other light fittings, but, physically they are particularly well suited to use in a workshop with a suspended ceiling. They need a hole cut in the ceiling to match the size of the unit. Manufacturer's instructions include details of the

Fig 10.16 *Spotlamps are one very effective way of providing adjustable task lighting. Single, and multiple units, as well as track mounted systems can all be obtained.*

Fig **10.17** *A downlighter, installed in the ceiling of a workshop, can illuminate a single fixed area; useful above a workbench.*

size of hole to be cut, and often a template to assist in the task.

Portable lights, such as anglepoise and flexible machinery lights are almost the perfect solution to providing the required lighting level over machinery where delicate work is undertaken, such as on lathes, milling machines, drill presses and bench grinders. Their big advantage is that the source of the light can be moved to match the exact task in hand, and shadows in the wrong place can usually be eliminated entirely. The use of reflector bulbs in place of the more conventional type considerably improves the illumination efficiency.

The type of anglepoise which is fitted in place with a clamp is to be preferred for two reasons. First, it can be fitted more securely to the workbench or machinery stand, and secondly, it does not require the space occupied by the large, heavy, round base of the free standing variety. Exactly the same guidelines apply for flexible machinery lights, but these are rather more expensive to purchase and are only available from specialist suppliers.

For those with less than perfect eyesight, or for those working on really small models or parts of models, a magnifying light can be a great asset. These lamps are built on a similar principle to the anglepoise, but incorporate a large magnifying lens with illumination around its periphery. The light is normally provided by a single incandescent bulb, but more expensive variants carry a circular fluorescent tube. Again for workshop use, those fitted with a clamp are to be preferred to free standing ones.

Supplying lights from a socket

Portable lights should not be permanently wired into a lighting circuit, but instead connected to a convenient 13 amp

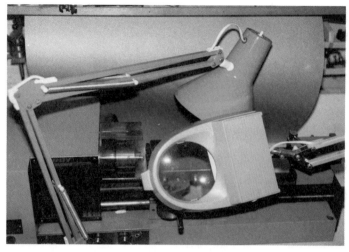

Fig 10.18 *Anglepoise and magnifying lights are practical for lighting a wide range of tasks, and can quickly be moved out of the way.*

socket. A standard 13 amp plug should be used and fitted with a 2 or 3 amp fuse; sufficient for 480 to 720 watts of lighting! Normal twin core flex is fine, but the route from the light to the plug must be well clear of all machinery.

Another danger is the possibility of tripping over loose cables. Two rules apply here. First, try and cut the cable to the right length before fitting the plug or use a coiled cable, which will automatically adjust to length. Second, if there aren't enough conveniently placed sockets, then install some more. Lighting is one of the few areas where the use of two or three way adapters is acceptable, as there is little likelihood of them becoming overloaded with the relatively low power consumption of light bulbs.

Inspection lights

A portable inspection light, with a wire cage to protect the bulb, and a long lead is a particularly useful piece of equip-

Fig 10.19 *An inspection light, an essential for car maintenance, will find many uses in the home workshop.*

ment in the home workshop. It is ideal for getting an adequate level of illumination into the interior of large items of machinery. Developed for the car repair industry, these lights are relatively inexpensive and widely available.

Emergency lighting

Sooner or later, workshop lighting, like any other, is likely to fail. It may be a power cut, it may be a blown fuse or tripped circuit breaker, but regardless, if it is dark, some sort of emergency lighting will be needed. Therefore, from a safety point of view, it is a good idea to keep a torch handy in the workshop, in a place where it can always be found. The ideal position is in a clip next to the fire extinguisher. Alternatively, specialist emergency light fittings can be purchased. These incorporate a trickle charged battery and an automatic changeover switch, which turns on the emergency light should the mains fail, and off again once power is restored. They may be wired directly from any lighting circuit and do not even require a switch. Such lights are not, however, a cheap solution to a hopefully rare problem.

Conclusion

Many and varied are the types of light fittings and bulbs. Ideas can be obtained by visiting big shops and stores, which carry large stocks of items, and new developments in the lighting field can be observed at first hand. There is really little excuse these days for having a badly lit workshop, apart from affordability. However, it is noticeable that many light fittings are thrown out, both by industrial and commercial users, as well as the average household, simply because they are no longer cosmetically acceptable. This provides an ideal source of low cost, if not free, secondhand units. With the running cost of lights small compared with the expense of operating machine tools and providing heating, a well lit workshop will reward its owner many times over.

CHAPTER 11

Three phase supplies

General

It is an interesting fact that the Electricity Boards run mains power into each street in three phase form, and only distribute it down to single phase at the distribution units connecting up individual premises. The three phases are provided on three wires, one for each phase, together with a neutral and an earth wire. Single phase electricity is fine for the supply of power levels up to 10.8 kW, a figure most unlikely to be exceeded in a home workshop, but above that, it is practically much simpler to provide electricity in its three phase form. Home workshops can normally be run quite happily from a single phase source, and, as the vast majority of homes are only wired for this form of power, equipment suppliers to the home workshop market have concentrated on machines designed to be run from single phase mains.

The advantages and disadvantages of three phase electricity

The advantages of three phase electrical supply are threefold. First, it is much easier to provide high power levels. This, in itself, is unlikely to be a major consideration for a home workshop. As power demands increase, with larger

machines, the other benefits of three phase supplies become apparent. Electric motors can be made significantly smaller and lighter, for a given power output, and three phase induction motors will self start without any further ancillaries. These advantages were recognised long ago by industry, and most machinery found in industrial premises operates from a three phase electrical supply.

The negative factors are that the voltage, at 415 volts, is potentially even more lethal than the 240 volts of the single phase mains, and that additional expense is involved in providing a three phase supply in the workshop. **It is most important to note that single phase equipment and lighting must never be connected directly to a three phase outlet.**

The reason why a model engineer might want to run three phase equipment in a home workshop is almost certainly because the vast majority of ex-industrial machines require such a supply, and this chapter concentrates on that particular reason. These machines are widely available on the secondhand market, and can often be picked up for bargain prices when the companies using them either re-equip or cease trading. It is considered unlikely that a new three phase machine would be chosen for home workshop use unless a three phase supply already existed there.

Obtaining a three phase supply

Whilst it might seem that the first choice for anybody needing a three phase supply would be to get the local Electricity Board to install one, this is not likely to be the easiest solution for the average home workshop. The cost will depend on a number of factors, the main one being the length of the cable run which has to be installed. The Board will normally provide a five wire system; three phases, neutral and earth, and these will have to be connected by the Board to the user's three phase distribution unit, which will provide circuits to all the machines which require it in the workshop. In addition, special three phase accessories, such as plugs, sockets and switches will be needed, together with suitable wiring.

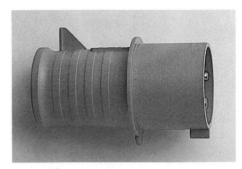

Fig 11.2 *A typical three phase plug, available in four or five pin versions, the former being ideal for use with three phase converters.* (Photo courtesy MK Electric.)

The provision of the neutral wire allows that wire to be used with any one of the phases to provide a single phase supply, which may be required for the lighting circuits, as well as any single phase machinery. However, as it is considered that a model engineer will not normally wish to install and use such a three phase supply in a home workshop, due to its expense and additional complexity, its provision is considered to be beyond the scope of this book.

Fig 11.1 *A three phase socket looks completely different from its single phase equivalent. This one includes a twist on/off switch.* (Photo courtesy MK Electric.)

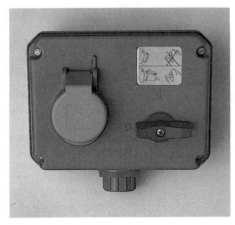

Three phase converters

The alternative is to purchase a converter to provide a three phase supply from a single phase source; a proposition which is both practical and affordable. These converters depend on the interaction between the converter and the motor being powered to induce an artificial three phase output. Reliable, economically priced units are available from a number of suppliers, and, whilst the higher powered versions almost invariably require a direct, high current, single phase feed circuit from the consumer unit, lower powered versions can be run from

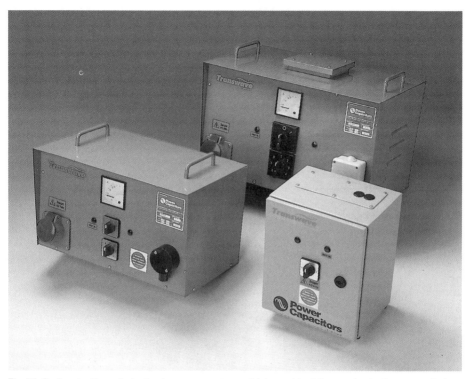

Fig 11.3 *A selection of single phase converters which are ideal for use in the home workshop.* (Photo courtesy Power Capacitors Ltd.)

a ring main circuit.

The simplest units enable just one machine to be supplied with three phase power. Rather more expensive versions are available, which allow several machines to be run. Units covering the range from 1.1 kW to 7.5 kW (1.5 HP to 10 HP) are the most common, and their efficiency ensures that little power is lost in the conversion to a three phase supply. It must be noted, however, that each converter has a minimum as well as a maximum loading for satisfactory operation. The reliability of these converters is excellent, providing the manufacturer's operating instructions are obeyed. The use of such a converter is clearly the preferred approach for the home workshop with one or perhaps two ex-industrial three phase machines.

Running three phase machines from a single phase power supply

If a decision has been made to run three phase machinery via the existing household single phase supply using a converter, then a number of factors need to be considered when planning the connection of the three phase machine tools. These are the power requirements of each machine, the number of electric motors to be run from the three phase supply, and whether they will be operated individually, or at the same time.

The most important factor is to match the converter output not only to the machine's power requirement but also to its starting current, which normally must not exceed three times the full load current rating of the motor. Converters automatically cope with the motor starting surge, until the motor has achieved its full running speed. Where the converter is required to power more than one motor, a power selector switch allows the operator to choose the optimum power to suit the particular motor or combination of motors to be powered. When a combination of three phase motors is supplied simultaneously, it is essential to ensure, not only that **the combined rating of the motors does not exceed that of the converter**, but also that **the most powerful motor is started first**. The power output from the converter is normally provided via a three phase four way socket.

Very occasionally, some small three phase motors, particularly those with unusual magnetic characteristics, for multi speed or reversing applications, as well as motors subjected to abnormal load characteristics, require a pilot motor in parallel to the driven motor to maintain a satisfactory third phase. The converter manufacturers can advise where this might be necessary, in which case the pilot motor should be at least equal in capacity, preferably larger than the driven motor. A four pole (1425 rpm) motor fitted with starter and overload protection is ideal.

Connecting to a three phase supply
Assuming that a three phase supply is to be provided by a three phase converter,

the installation and connection of the machine(s) is quite straightforward. The supplier of the converter will include comprehensive fitting instructions, and, for wiring to a single machine, direct connection to the converter is recommended using a three phase four-way plug. Where more than one machine is involved, provision must be made to connect equipment via a three phase distribution board. A fixed connection unit will also be needed of the appropriate rating to connect the converter to the single phase mains supply.

Wiring and distribution boards
The wiring colours for three phase supplies are as follows:

Phase R	Red
Phase Y	Yellow
Phase B	Blue
Neutral	Black
Earth	Green and yellow

The most important practical difference to remember between single phase supplies and the three phase supply from a converter is that the latter have a fourth wire; one for each of the three phases plus earth. Neutral is normally only provided with Electricity Board three phase supplies, although some converters are available with five wire connections. Such converters are not needed for home workshop use. Concentrating on four wire connection, as with single phase wiring, **the cable size and fuses must be selected to suit the current required**. Switch units and junction boxes are all different, and supplied with a fourth way. In addition, with 415 volts being used, the demand on insulation is greater, but this is taken care of by the manufacturers of the various fittings. These differences apart, all the basic

Fig 11.4 *This triple pole switch is ideal for three phase machine tools, capable of handling 32 amps at up to 440 volts.* (Photo courtesy MK Electric.)

rules of single phase wiring still apply, and wiring up for three phase supply in the workshop is hardly more difficult than for normal single phase mains.

Starting and speed control

One of the advantages of three phase motors mentioned earlier is their ability to self start without any additional equipment. Thus, no starting capacitors are needed, nor is there any internal switch gear. The control of the speed of three phase induction motors, like their single phase brethren, is virtually impossible, although specially built controllable motors, with power output up to a few watts, can sometimes be found.

CHAPTER 12

Low voltage supplies

General

The definition of low voltage supplies for the purposes of this book is any voltage below 50 volts. The reason for this definition is that below 50 volts, an electric shock is considered non-lethal. However, when a low voltage supply is derived from a transformer, and it usually is, then **the transformer must be a safety isolating one**. It must comply with BS 3535, with no connection between the output winding and the body of the transformer or the earthing circuit. The vast majority of transformers meet this requirement, but so-called auto transformers do not include such isolation.

The main uses of low voltage supplies are for driving 'on lathe' spindles, for power feeds for tables, and for suds pumps, together with their employment for low voltage lighting systems for machine tools. Their advantage is that the proneness of the associated cables to damage does not bring with it the danger of serious electric shock. This, together with the small size and wide choice of low voltage motors available, both new and secondhand at affordable prices, makes them a natural choice for such applications.

Low voltage lighting

Low voltage lighting systems can be purchased from a number of high street sources, and make use of high efficiency quartz halogen bulbs. The use of fittings with small reflectors makes them ideal for providing task lighting for the working area of any machine tool, and, whilst the cable from the light must be run clear of the machine tool, any accident with the cable will be an inconvenience rather than potentially lethal.

There are, however, penalties to pay for the advantages of low voltage lighting. First, the correct transformer for the number of light fittings is important. Overloading the transformer will under volt the lights, resulting in reduced light output. In addition, there is the risk of overheating the transformer. Equally, underloading the transformer will tend to over volt the lights, resulting in markedly reduced bulb life. In addition, because of the low voltage, the current will be high. For example, a 60 watt bulb will draw 5 amps at 12 volts. Thus, larger cable sizes will be needed for these lights than for their 240 volt brethren.

12 volt equipment

Because of the wide availability of 12 volt

components for the automobile industry, practical examples will be limited to this voltage, but 24 volt equipment is also reasonably available and does have some benefits as it halves the current requirement for a given power output. 6 volt motors are widely used in model radio controlled cars, but offer no advantages which would justify their use in a home workshop, unless such R/C cars are a hobby!

So far in this book, all electrical supplies have been alternating current ones. For low voltage systems, both alternating current (AC) and direct current (DC) systems can be used. The various pros and cons need to be understood. For low voltage lighting, AC is perfectly satisfactory, and this can be obtained from nothing more complicated than a transformer of suitable voltage and current ratings. For 12 volt systems, a 40 watt lamp will consume over 3 amps, so that, with several lights in use at the same time, a fairly beefy transformer will be needed. Such a unit can be obtained from a low voltage lighting supplier or from an electronic component supplier. Low voltage AC motors, whilst perfectly satisfactory in theory, are uncommon in practice, and are unlikely to be found in the average home workshop.

12 volt power supplies and speed controllers

Power supplies for DC systems are somewhat more complex than AC ones, requiring, as well as a transformer, a bridge rectifier, smoothing capacitor and, for preference, an individual fuse. Such units are available as ready-made power supplies and their construction is also within the capability of most model engineers. An alternative, favoured by some who require a mobile solution, is to use a car or caravan 12 volt battery to provide the prime 12 volt power source, supplemented with a battery charging system. Even the addition of speed control is well within the building capabilities of

Fig 12.1 *12 volt DC power supplies are widely available. This one is designed to power small hand-held power tools.*

the amateur, but again, there is a fair range of off-the-shelf devices available on the market. High quality speed controllers are reasonable inexpensive to purchase and, for 12 volt systems in particular, stock items have been developed for the model railway and electric slot racing car markets. These can readily be adapted for home workshop use.

12 volt wiring
As with all the other forms of flexible wiring described in this book, low voltage DC wiring should normally be colour coded; either red for positive and black for negative or black with a white stripe for positive and black for negative. However, for fixed installations, bell wire, with its twin white insulated copper conductors is quite satisfactory for low current carrying circuits.

12 volt tools
There are now a number of manufacturers marketing a range of small low voltage power tools, such as drills, saws, engravers and sanders, and they also market suitable power supplies, often including built-in electronic speed control. Such tools are increasingly to be found in the home workshop, and their adaptation to use with machine tools is remarkably simple. Airbrush compressors, so handy for getting a perfect paint finish on models, are also generally powered from a 12 volt supply.

For coolant pumping, such low voltage systems have a double advantage, due to the enhanced dangers of mains electricity in the presence of water. Where coolant systems are supplied with the machine tool, they are usually run from the power supply for the tool itself. However, where a coolant system is fitted by the model engineer after the machine has been purchased, a 12 volt

electric pump is the obvious choice for circulating the coolant. Such a system requires only a 12 volt power supply and switch to provide an electrical system which is safe to operate; suitable switches being available from car accessory shops.

Telephones
Telephone extensions are a step away from normal electrical systems, but are included here for completeness. They are particularly useful if the person using the workshop is the only person available to answer the telephone, particularly if the workshop is some distance from the house. Unless a shared party line is involved, in which case BT advice should be sought, adding an extension to the workshop is simplicity itself, assuming the home already has a BT (or Mercury) telephone socket fitted. Both telephones and installation kits are widely available for this task and contain detailed instructions. What is needed is a telephone, an extension socket, a suitable length of standard four core telephone cable with a BT telephone plug, and a telephone two way adapter. The adapter allows both the existing telephone and the extension to be connected to the BT socket when the work has been completed. The cable is a special four core one, which must meet BT specification CW 1308, and the connection to terminals is as follows:

Terminal 2	Blue with white markings
Terminal 3	Orange with white markings
Terminal 4	White with orange markings
Terminal 5	White with blue markings

There must not be more than 50 metres of cable between the master socket and any extension, or more than 100 metres of cable overall in a complete

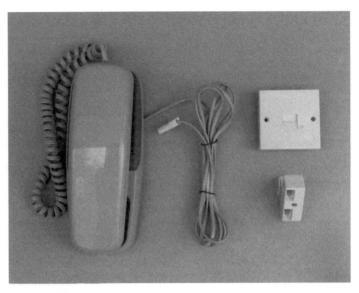

Fig 12.2 *A small telephone, telephone socket and two way adapter make the installation of a telephone in the workshop a simple and economic task.*

household installation. Telephone cabling must be kept at least 50 mm from mains electrical cable unless a special conduit is being used, when the telephone cable must be separated by a divider from the mains cabling. The plug end of the extension cable should be placed adjacent to the existing socket, with a small amount of slack to allow it to be plugged in. The cable should then run tidily along the wall, and through 6 mm holes drilled as necessary, being held in place with small plastic clamps. Whilst the cable may be run through roof or floor spaces, normal practice is, whenever possible, to run it along the top of skirting boards and around door frames. At the far end, the cable is wired into the socket, which, as with electrical sockets, may be surface or flush mounted. Either screw terminals are used in sockets, or V slot connectors for each wire; the wire being inserted with a special insertion tool supplied with the socket. Usually, the telephone

socket plate, which is somewhat smaller than an electrical socket, is fixed on a small surface mounting box, which gives an acceptable appearance. It is generally a good idea to mount the telephone in a workshop on the wall, to avoid potential damage to it.

Intercoms

Intercoms come in two kinds. There are those which can have the two terminals plugged into any convenient 13 amp electrical socket, and use the mains wiring for communication as well as power, and those which require low voltage wiring to be installed. The first type is so simple to install that no further information is needed other than to follow the maker's instructions. The low voltage type is normally battery powered, and need a two core bell cable run between the two terminals. Not a difficult task, this cable can be run as if it were standard telephone wire. The manufac-

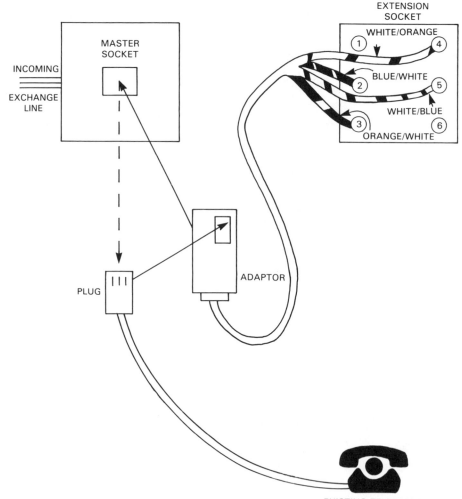

EXTENSION
SOCKET

MASTER
SOCKET

INCOMING
EXCHANGE
LINE

WHITE/ORANGE

BLUE/WHITE

WHITE/BLUE

ORANGE/WHITE

ADAPTOR

PLUG

EXISTING TELEPHONE

Fig 12.3 *Wiring a telephone extension to an existing socket is quite a straightforward task, with four wires to be connected into the extension socket. It is essential to ensure that the correctly coloured wires are attached to the right terminals.*

94

turer's installation instructions will give clear details. If the workshop is an outdoor one, any telephone or intercom cable will need to be routed either through an underground conduit or suspended by a catenary system.

Security systems

It is a sad fact of life that burglary from home workshops is a real risk, and a suitable security system can protect the contents as well as providing a deterrent effect. Such systems vary in complexity enormously. If there is an existing household security system, it is not a difficult task to lay additional wiring to the workshop to hook up sensors on the doors and windows. Infra red and ultrasonic movement detectors are rather more complex to install, as they require power as well as signal cables.

It may indeed be that, with an outside workshop, a self-contained security system will be considered a worthwhile investment. In such a case, a control panel, back-up power supply and alarm bell/siren will also be required. Whilst many may feel that all of this is the job for a professional security company, a number of DIY security kits are on the market, and these are as easy to fit as other electrical circuits. Using bell cable to interconnect the various sensors to the control unit, the whole system is usually powered, via an internal mains transformer, from a fixed connection unit. Manufacturer's instructions for DIY kits are generally extremely explicit.

Fig 12.4 *Security systems can vary widely in cost and complexity. Serious consideration should be given to the protection of the valuable contents of the average home workshop.*

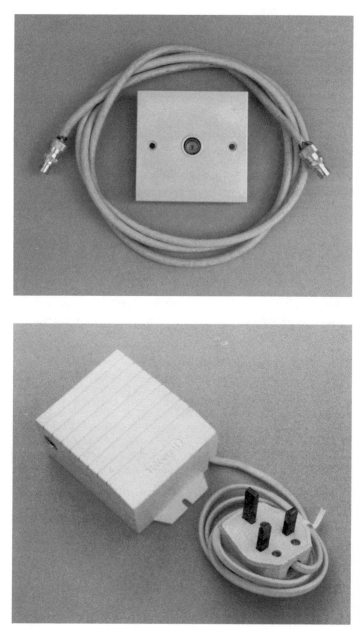

Fig 12.5 *A television in the workshop can provide useful companionship, and an aerial socket is readily installed and connected to the TV using co - axial cable.*

Fig 12.6 *TV signal boosters are available, which are mains powered, and are connected between the aerial and TV set to improve reception.*

TV aerials

A television in the den, as a source of company, is not uncommon. Whilst the set can simply be plugged into an adjacent power socket and used with an indoor aerial, in many parts of the country, a proper outdoor aerial is needed. Two approaches to this problem are considered. First, it is straightforward to install an outside antenna on the roof of the workshop, and wire it to an aerial socket using co-axial cable. Such sockets are the same size as 13 amp electrical sockets and can be surface or flush mounted. Alignment of the aerial should start by pointing the aerial in the same direction as others in the neighbourhood, with the antenna itself mounted with all the elements in a horizontal plane. Fine adjustments can then be made, with a helper looking at the picture quality on all channels. There may sometimes need to be a compromise between the best picture on one particular channel and an acceptable picture on the rest.

The alternative is to run an extension from an existing socket. This will require the co-axial cable to be laid as if it were electrical cable, and may require a booster at the television if the signal is weak. These boosters are powered from a 13 amp socket. They have an input for connection to the TV aerial and an output, which requires a short length of co-axial cable, for connection to the set.

Glossary of terms

This glossary includes an abridged list of the definitions given in the *IEE Wiring Regulations* 16th edition, some of which are worded very legalistically, together with a number of other useful terms and abbreviations.

AC   Alternating current.

Accessory   An electrical item, such as a switch, socket or fixed connection unit, permanently connected to a circuit.

Adapter   A device used to allow more than one appliance to be plugged into a socket.

Ampère or amp   The unit of measure of the flow of electrical current.

Appliance   An item of electrical equipment other than a light or independent motor.

Bridge rectifier   A device for converting alternating current to direct current.

Cable   One or more electrical conductors in an outer sheathing, separately insulated or otherwise protected by the outer covering.

Capacitor   A device for storing small quantities of electrical energy.

Cartridge fuse   A device comprising a fuse element(s), enclosed in a cartridge usually filled with arc-extinguishing medium and connected to terminations.

Catenary   A system for supporting any overhead electrical cable not routed through a conduit.

Ceiling rose   A customised type of junction box used for connecting a pendant light to its circuit.

Ceiling switch   A light switch, which attaches to the ceiling and is operated by pulling a cord.

Choke   A coil of wire wound around an iron core, found in some fluorescent tube fittings.

Circuit   An assembly of electrical equipment supplied from the same origin and protected against overcurrent by the same protective device.

Circuit breaker   A device capable of making, carrying and breaking normal load currents and making and automatically breaking, under predetermined conditions, abnormal currents, such as short circuit currents. It is usually required to operate only infrequently.

Conductor   A length of electrical wire along which an electric current can flow.

Conduit   A part of a closed wiring system for cables in electrical installations, allowing the cables to be drawn in and/or replaced, but not inserted laterally.

Connector The part of a cable coupler or of an appliance coupler which is provided with female contacts and is intended to be attached to the end of the flexible cable remote from the supply.

Consumer unit A unit, owned by the householder, which is connected to the Electricity Board's supply, and which provides the user with a means of isolating the supply, as well as distributing it through a number of fuse or MCB protected circuits. It is one form of distribution board.

Current The quantity of electricity flowing through a conductor. It is measured in ampères (amps) and milliamps.

Current-carrying capacity of a conductor The maximum current which can be carried by a conductor under specified conditions without its steady state temperature exceeding a specified value.

DC Direct current.

Distribution board An assembly containing switching and protective devices (e.g. fuses or circuit breakers) associated with one or more outgoing circuits fed from one or more incoming circuits, together with terminals for the neutral and protective circuit conductors. Means of isolation may be included in the board or may be provided separately.

Double insulation Insulation comprising both basic insulation and supplementary insulation.

Double pole switch A switch which breaks both the phase and neutral connections.

Downlighter A light which fits flush in the ceiling, directing a narrow pool of light downwards.

Duct A closed passage way formed underground or in a building and intended to receive one or more cables which may be drawn into the duct.

Earth (E) The conductive mass of the Earth, whose electric potential at any point is conventionally taken as zero. Earth wiring must be connected to the Earth.

Earth leakage current A current which flows to Earth in a circuit which is electrically sound.

ELCB An earth leakage circuit breaker, now obsolete, and replaced by residual current devices (RCDs).

Electric shock A dangerous physiological effect resulting from the passing of an electrical current through a human body.

Emergency stopping Emergency switching intended to stop a dangerous movement of a machine.

Extension lead A length of flex with a plug on one end and a socket on the other, allowing an appliance to be connected to a distant socket.

Fixed connection unit A device used to connect an appliance permanently into a ring main circuit.

Flex Wiring designed to provide mechanical flexibility without degradation of the electrical components.

Fuse A device that by the fusing of its specially designed component, opens the circuit in which it is inserted by breaking the current when this exceeds a given value for a sufficient time.

Grommet A ring of rubber/plastic used to line a hole in metal through which an electrical cable passes, to prevent chafing.

Insulation Suitable non - conductive material enclosing, supporting or surrounding a conductor.

Isolation A function intended to cut off, for reasons of safety, the supply from all, or a discrete section of the installation by separating the installation or section from every source of electrical energy.

Kilowatt   One thousand watts.

Kilowatt hour   A consumption of one kilowatt for one hour, 100 watts for ten hours or ten kilowatts for six minutes.

Live conductor (L)   An obsolete term for the conductor of an AC system for the transmission of electrical energy, other than a neutral or a protective conductor. It is now called the phase conductor.

Live part   A conductor or conductive part intended to be energised in normal use, including a neutral conductor.

Loop-in rose   A ceiling rose which combines the function of a junction box, allowing both the light and the switch to be connected to the mains.

Mains   An abbreviation for the 240 volt, single phase, alternating current electricity provided by the Electricity Boards.

MCB   A miniature circuit breaker which opens the circuit in which it is inserted by breaking the current when this exceeds a given value for a sufficient time.

Megohm   One million ohms.

Milliamp   One thousanth of an amp.

Neutral conductor (N)   A conductor connected to the neutral part of a system and contributing to the transmission of electrical energy.

Night storage heater   A heater which stores heat generated by low tariff electricity overnight and then releases the heat the next day.

Ohm   The measure of electrical resistance.

Phase conductor (P)   A conductor of an AC system for the transmission of electrical energy, other than a neutral conductor, or a protective conductor.

This was previously known as the live conductor.

Plug   A device, provided with contact pins, which is intended to be attached to a flexible cable, and which can be plugged into a socket.

Portable equipment   Electrical equipment which is moved while in operation or which can easily be moved from one place to another while connected to the supply.

Protective conductor   A conductor used for some measure of protection against electric shock and intended for connecting together any of the following parts:

The main earthing terminal
Earth electrode(s)
Exposed conductive parts

Radial circuit   A circuit which distributes electrical current along a single cable from the point of supply to an appliance.

Residual current device (RCD)   A mechanical switching device, intended to cause the opening of the contacts when the residual current attains a given value under specified conditions. It is used to prevent electric shocks.

Ring main circuit   A circuit, arranged in the form of a ring and connected to a single point of supply, directly connected to appliances, or to a socket(s) or other outlet points for connection to appliances.

Sheathing   The outside insulation of a cable, containing two or more insulated conductors, with or without a single uninsulated conductor.

Short circuit current   An overcurrent resulting from a fault of negligible impedance between live conductors having a difference in potential under normal operating conditions.

Socket   A device, provided with female contacts, which is intended to be installed with the final wiring, and intended to receive a plug.

Spotlight A light fitting which allows its beam of light to be pointed in a selected direction.

Spur A branch from a ring main circuit.

Switch A mechanical device capable of making, carrying and breaking current under normal circuit conditions, which may include specified overload conditions, and also of carrying for a specified time currents under specified abnormal conditions such as those of short-circuit.

Terminal A connection for an electrical conductor.

Transformer A device for altering the voltage of an alternating current supply.

Voltage, nominal Voltage by which an installation, (or part of an installation) is designated. The following ranges of nominal voltages (rms values for AC) are those likely to be found in the home workshop, and are defined as:

Extra low Normally not exceeding 50 volts AC or 120 volts ripple free DC, whether between conductors or to earth.

Low Normally exceeding extra-low voltage but not exceeding 1000 volts AC or 1500 volts DC between conductors or 600 volts AC or 900 volts DC between conductors and earth.

The actual voltage of the installation may differ from the nominal value by a quantity within normal tolerances.

Watt The measure of electrical consumption of one or more items of electrical equipment.

List of useful addresses

Electromail,
PO Box 33,
Corby,
Northants.,
NN17 9EL.

Institution of Electrical Engineers,
PO Box 96,
City House,
Stevenage,
Herts., SG1 2SD.

Maplin Electronics,
PO Box 3,
Rayleigh,
Essex, SS6 2BR.

Martindale Electric Ltd.,
Neasden Lane,
London,
NW10 1RN.

MK Electric Ltd.,
Shrubbery Road,
Edmonton,
London, N9 0PB.

Motorun Phase Converters,
23, Waldegrave Road,
Teddington,
Middx.

Osram Ltd.,
PO Box 17,
East Lane,
Wembley,
Middx., HA9 7PG.

Philips Lighting Ltd.,
PO Box 298,
420/430 London Road,
Croydon,
Surrey, CR9 3QR.

Power Capacitors Ltd.,
30, Redfern Road,
Tyseley,
Birmingham, B11 2BH.

RS Components Ltd.,
PO Box 99,
Corby,
Northants, NN17 9RS.

Index

Adapter 18, 33, 36, 38, 83, 98
Armoured cable 41, 49, 50, 53, 62

Battery charger 5, 21, 56, 91
Bell wire/cable 48, 92, 93, 95
British Standards 9, 14
Brush motors/gear 13, 31

Cartridge (fuse) 33, 34, 35, 37, 38, 62, 98
Catenary 51, 95, 98
Circuit breaker 84, 98, 99, 100
Co-axial cable 48, 97
Conduit 51, 93, 95, 98
Continuity tester 12, 38, 48
Converters 1, 86, 87, 88

DC 5, 48, 91, 92, 99, 101
Distribution unit 26, 27, 85, 86
Double insulation 19, 59, 99
Downlighter 99

ELCB 17, 35, 99
Electrician 1, 8, 13, 24, 25, 26, 48
Electrocution 15, 17, 23, 35

Filter 31
Fire extinguisher 24, 84
First aid 23
Flex connector 18, 21, 25, 63

Garden 3, 17, 37
Generator 7, 31
Gland 50, 53
Grinder 5, 6, 56, 82
Grommet 46, 60, 99

Hertz 5, 31
High power 5, 6, 16, 46, 67, 85
HP 5, 6, 87

IEE 8, 9, 13, 47, 98
Insulating tape 21, 46, 63
Intercom 11, 93, 95
Interference 13, 31, 48
Isolating 15, 37, 53, 90, 99

Junction box 26, 41, 46, 49, 50, 88, 98, 100
Junction box (lighting) 70, 71, 74, 77

Kilowatt 4, 5, 6, 7, 26, 34, 35, 56, 62, 100

Lathe 4, 5, 9, 56, 68, 82, 90
Low voltage 1, 10, 14, 19, 32, 48, 70, 90

Machinery 82−87
MCB 16, 17, 25, 28, 31, 33−37, 53, 56, 67, 71, 99, 100

Medical   23, 25
MICS   41, 49, 50, 53
Milling machine   5, 9, 56, 68, 82
Multimeter   12, 38

Night storage   6, 7, 26, 35, 40, 69, 100

Off peak   7, 10, 26
Ohm   4, 38, 48, 100
Overhead   49, 51, 98
Overloaded   23, 39, 83

Permanent wiring   16, 40, 42, 46, 56,
   82
Power supply   6, 9, 49, 92, 95
Pull cord switch   74, 98
PVC   40–42, 44, 49–51, 53, 69, 70

Quick start   77, 79

Radial circuit   8, 40, 44, 46, 49, 53,
   67, 69, 70, 100
Radio   4, 5, 13, 31, 35, 56, 62, 64, 91
RCD   25, 33, 35–37, 48, 53, 54, 59,
   99, 100
Regulations   2, 8, 9, 13, 47, 48, 98
Resistance   4, 48, 76, 100
Rubber   46, 61, 99

Saw   6, 9, 12, 13, 43, 56, 68, 92
Security   1, 6, 11, 49, 54, 55, 74, 95
Speed control   31, 64, 65, 88–92
Spur   6, 40, 53, 57, 60, 101
Standby power supplies   31
Starter   4, 9, 77, 79, 88
Suds pump   5, 90

Tariff   6, 8, 26, 69, 100
Task lighting   11, 70, 80, 81, 90
Telephone   11, 48, 92, 93, 95
Termination   53, 98
Test   9, 12, 13, 25, 29, 37, 38, 47, 48
Test equipment   6, 13, 46, 48, 56
Tester   12, 13, 38, 48
Thermostat   64, 68
Transformer   31, 77, 90, 91, 95, 101
Trip   17, 35–37, 84
TV   1, 5, 35, 56, 62, 97
Two way   40, 75, 92

US   5, 31

Ventilation   6

Water, dangers of   15, 24, 92
Water analogy   4, 5
Waterproof   53, 54, 59